Space, Time, and Gravity

Space, Time, and Gravity

The Theory of the Big Bang and Black Holes

Robert M. Wald

The University of Chicago Press
Chicago and London

The University of Chicago Press, Chicago 60637
The University of Chicago Press, Ltd., London

© 1977 by The University of Chicago
All rights reserved. Published 1977
Printed in the United States of America
82 81 80 79 78   98765432

Robert M. Wald is assistant professor in the
Enrico Fermi Institute and the Department
of Physics at the University of Chicago.

Library of Congress Cataloging in Publication Data

Wald, Robert M
Space, time, and gravity.

Bibliography: p.
Includes index.
1.   Cosmology—Addresses, essays, lectures.
2.   Space and time—Addresses, essays, lectures.
3.   Gravitation—Addresses, essays, lectures.   I.   Title
QB981.W24   523.1   77–4038
ISBN 0–226–87030–8

Contents

Preface

This book arose directly from a series of ten public lectures which I gave at the University of Chicago during the spring of 1976. This lecture series— entitled the Compton Lectures in honor of Arthur H. Compton— was established in 1976 with funds from the bequest of John W. Watzek, Jr., for the purpose of giving interested persons both within and outside the University community the opportunity to find out what is happening in current scientific research, especially in those areas being pursued by the Enrico Fermi Institute. Each of the chapters in this book corresponds to one of the lectures in the spring 1976 series.

The subject of this book is our present-day ideas of space, time, and gravitation. The first three chapters lay the foundations by explaining how, in Einstein's theory of general relativity, gravitation is described in terms of the curved geometry of spacetime. The next two chapters discuss the implication of these ideas for the universe: its origin, evolution, and large-scale structure. The last five chapters are devoted to the subject of gravitational collapse and black holes. Many exciting developments have taken place recently in these areas; indeed, the theory of black holes has only been fully developed within the past decade. Many of the ideas discussed in the book are presently very active areas of scientific research.

I have tried to write the book in such a way that all the important and basic ideas are described in depth in a clear, logical manner, without oversimplification and without inclusion of irrelevant, technical detail. If I have succeeded, this book should be understandable and enjoyable to any reader who is curious about the ideas discussed here, largely independent of his previous scientific or mathematical background.

It is a pleasure to acknowledge a number of conversations with John Friedman, Robert Geroch, and David Malament which helped improve my presentation of special relativity (chap. 2). I wish to thank Jean Early and my wife, Veronica, for typing the manuscript. I also wish to thank Veronica for making a number of helpful criticisms of the book from the viewpoint of the layman. Most of all, I wish to express my appreciation to the Enrico Fermi Institute of the University of Chicago for appointing me as the second Compton lecturer and providing me with other assistance in the writing of this book.

One

The Geometry of Space and Time

1.1 Introduction

More than two thousand years ago, Euclid wrote down his postulates and proved a number of theorems concerning the geometry of space. His work is still used as a model of mathematical reasoning. Until this century, mathematicians like Euclid and philosophers have dominated all discussions of space and time. However, the attitude which will be adopted throughout this book is that the investigation of the properties of space and time properly belongs to the domain of physics. The question of whether or not Euclid's postulates are satisfied in the real world is properly settled by doing experiments and making observations. All our day-to-day experience indicates that Euclid's postulates do indeed hold, but we should leave our minds open to the possibility that when we make careful observations on large distance scales, we may find that Euclid's postulates are violated.

In the first three chapters of the book we shall focus on the following basic question, What are the intrinsic properties of space and time? By this I mean, what are the properties which we can properly ascribe to space and time itself, without reference either to objects or fields which may happen to be present or to particular observers who make the observations? In this chapter, I shall begin the investigation by stating a number of key assumptions about space and time and deriving some conclusions from them. Although the language I will use in the discussion may seem somewhat unnatural, everything I will say in this chapter should be in complete accord with—or at least, not run contrary to—the experience and intuition of people who are not familiar with the theory of relativity. However, we shall find that of the four key assumptions we shall state explicitly, one of them (assumption (*ii*) below) is simply false (see chap. 2) while two of the others (assumptions (*iii*) and (*iv*)) no longer hold when the effects of gravity are taken into account (see chap. 3). Indeed, the purpose of this chapter is to state the commonly held ideas about space and time in a language which makes it possible to say what is wrong with them and how they should be modified, as we shall do in the next two chapters.

The first concept we need to introduce is that of an *event*. In common usage, an event is something which occurs at a certain place at a certain time. Here any point of space at any instant of time (even if nothing interesting is happening there) will be referred to as an *event*. One can think of space and time as being composed of events; the collection of all events encompasses all of space and time.

Rather than continually repeating the phrase *space and time,* I shall from now on use the single word *spacetime* to denote the collection of all events. By making this word substitution, I do not mean to imply at this stage that one must think of space and time as inseparable concepts, though in the next chapter we shall find that this is indeed true.

The first important fact about spacetime is that it is four-dimensional. To some readers, it may seem that in saying this I have already violated my promise to say nothing in this chapter which is contrary to common experience and intuition. But, in reality, the statement that spacetime is four-dimensional is a mundane fact (though a very important one). It simply means that in order to specify an event, one must give four numbers—one for the time, and three for the spatial position. This fact is familiar to anyone who has ever made an appointment (although frequently one only gives two numbers to describe spatial position, for example, numbers of the street and cross-street, it being understood that the meeting is to take place on the surface of the earth).

Before continuing our discussion of space and time, I wish to make two brief diversions: one to describe the very useful tool of spacetime diagrams (1.2) and the other to discuss the geometry of a sphere (1.3).

1.2   Spacetime Diagrams

In order to describe things that occur in spacetime, it is very convenient to draw pictures, called spacetime diagrams, in which a sequence of events in spacetime are depicted. The conventions on these diagrams are as follows: the horizontal direction denotes space, while the vertical direction denotes time, with "forward in time" being "upward" in the diagram.

To interpret a spacetime diagram, begin by looking at a horizontal section of it at the bottom of the diagram. This represents the spatial configuration of the objects at the beginning of the sequence of events depicted by the diagram. Horizontal sections which are higher up in the diagram give the spatial position at a later time. Thus, by moving one's eye upward on a spacetime diagram one can read off the motion of all the depicted objects. The path of an object in a spacetime diagram is often referred to as a *world line*.

There is one serious shortcoming of spacetime diagrams. As discussed in the previous section, spacetime is four-dimensional, but the piece of paper at one's disposal is only two-dimensional. By use of perspective, one can illustrate two spatial dimensions along with one time dimension, but we are still one dimension short. Thus, at least one spatial dimension is always suppressed in a spacetime diagram; therefore, one must use care in drawing and interpreting them.

Figures 1 through 4 are examples of spacetime diagrams. It is suggested that the reader get some practice by interpreting them and drawing spacetime diagrams depicting other sequences of events.

1.3   The Geometry of a Sphere

In this section, I wish to raise and answer a question about the sphere (for example, the surface of the earth) which is completely analogous

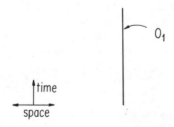

Figure 1   A spacetime diagram depicting the world line of an observer O_1, who is "standing still."

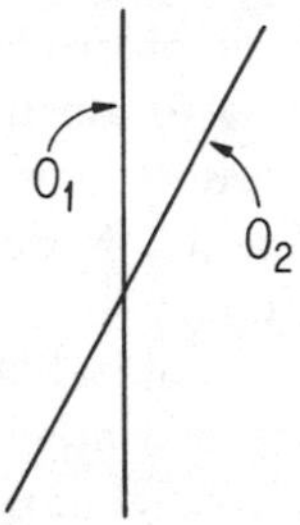

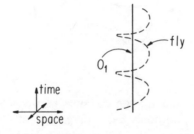

Figure 2   A spacetime diagram of O_1 and another observer O_2 who is moving past O_1 with uniform velocity.

Figure 3   A spacetime diagram of observer O_1 with a fly circling about him. (The fly *circles* around O_1; the world line of the fly *spirals* around the world line of O_1.)

Figure 4   A spacetime diagram of the wave front produced by a drop of water hitting a pond at event A. (The wavefront spreads out from A in a circular manner with uniform velocity.)

to the basic question about spacetime stated at the beginning of this chapter, What are the intrinsic properties of a sphere? The purpose of this section is twofold: (1) to illustrate the type of answer I am seeking to the analogous question concerning spacetime, and (2) to introduce a number of geometrical concepts which will prove very useful in the next two chapters.

The first statement one can make about the sphere is that it is two-dimensional, that is, it takes two numbers to specify a point on a sphere. These numbers are most commonly taken to be the latitude ψ and longitude ϕ of a point.

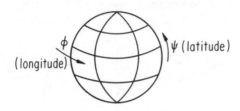

Figure 5   A sphere.

The *metric* is the formula for the distance between two nearby points.[1] In the notation commonly used, the metric of the sphere is

$$ds^2 = R^2 [(d\psi)^2 + \cos^2 \psi \, (d\phi)^2].$$

Here, ds denotes the distance between two nearby points, $d\psi$ their difference in latitude (expressed in radians), $d\phi$ their difference in longitude (expressed in radians), and R the radius of the sphere. Thus, using the above formula we can calculate directly the distance between two nearby cities, for example, Baltimore and Washington, if we know the latitude and longitude of these cities and the radius of the earth.

---

1. The mathematically minded reader undoubtedly will be unhappy with this definition. Exactly what do I mean by "nearby points"? Suffice it to say here that one can formulate a precise mathematical definition of the metric which captures the above intuitive notion. This definition, as well as mathematically precise statements of other geometrical notions, can be found in any modern textbook on differential geometry.

In fact, the metric contains all the information about the geometry of the sphere. For example, if we want to know the distance along any route between two faraway points (say, New York and Moscow), we can do so as follows: View the long route as a sequence of shorts routes, use the metric to calculate the length of these short routes, and add them up to get the length of the long route.

A *geodesic* is a line which curves as little as possible.[2] For the sphere, geodesics are also referred to as great circles; the equator is an example of a great circle. It is possible to show that the shortest path connecting any two points must be a geodesic. Thus, it is economical for airlines to follow great circle routes (geodesics) between cities. The concepts of the metric and geodesics will play a key role in the analysis of spacetime which we shall give in the next two chapters.

Finally, I wish to propose an answer to the question, What are the intrinsic properties of the sphere? The geometry of the sphere characterizes its intrinsic properties.[3] All the information concerning the geometry of the sphere is contained in the metric.

1.4   The Principle of Relativity

Returning to the subject of space and time, we now shall state our first key assumption.

Assumption (i) (Principle of Relativity): Of all possible motions of observers in spacetime, it is possible to distinguish *in an absolute sense* (that is, without making reference to the motion of other objects) a certain class of motions, usually referred to as "non-accelerating" or "inertial." Within this class there is no *absolute* way of distinguishing a preferred motion (for example, no inertial observer can be said to be "at rest" in an absolute sense).

In giving examples of this and other principles, I would prefer, until chapter 3, to avoid dealing with the issue of gravitational fields. Therefore, picture an astronaut in a rocket ship in outer space, far from other bodies or fields. The first half of the principle of relativity

---

2. See the previous footnote.

3. For the purposes of this discussion, I am ignoring topological characterizations.

states that, without looking outside the window of his rocket ship, the astronaut can tell whether or not his rocket engines are on, and what his acceleration is. The astronaut is said to be undergoing inertial moton if the engines are off. Although an inertial astronaut knows in an absolute sense that his acceleration is zero, the second half of the principle of relativity states that he has no way of determining his velocity except by a procedure tantamount to looking out of his window at other objects. The astronaut can "feel" the acceleration of the rocket ship—he will be pushed back in his seat and will feel very uncomfortable if the acceleration is large—but he cannot "feel" the velocity of the ship.

For the benefit of the reader who may have difficulty extrapolating experiences to that of an astronaut in outer space, let me point out that one can observe the same phenomenon every time one rides in a car. One can "feel" the acceleration of the car when it starts, stops, or turns. But, if there were no bumps on the road (which also cause the car to accelerate) it would "feel" exactly the same to stand still as to go at a uniform speed of 90 miles per hour on a straight road.

1.5   Absolute Simultaneity

The next assumption concerning spacetime is so deeply ingrained in the way most people think about spacetime that it is even difficult to be consciously aware of it. It is considerably more difficult to consciously trace out the logical consequences of this assumption and to see what changes must be made when we discover that this assumption is, in fact, false. This is why the special theory of relativity (see chap. 2) may, at first, seem quite strange and even paradoxical.

Assumption (ii) (false): "There is no limit to how fast an observer or other material object can move." More precisely, the events which cannot, in principle, be connected to a given event by the motion of an observer are three-dimensional ("space").

I might have stated this assumption in the following manner: "Given any two events, it is always possible, in principle, for a single observer to be present at both events unless these events occur at exactly the same time." The problem with stating the assumption this way is that the phrase "at the same time" already presupposes the main content of assumption (*ii*). In fact, as we shall see in the next chapter, when assumption (*ii*) is corrected, the notion of simultaneity ("at the same time") will be an observer-dependent concept. This illustrates some of the difficulties referred to in the first paragraph of this section.

Figure 6 illustrates the content of assumption (*ii*). From a given event A, it is possible for an observer to go to many later events, such as event B. It is also possible to get to A from many earlier events, such as C. By assumption (*ii*) the sets of events such as D having the property that it is impossible to go from D to A or from A to D form a three-dimensional "surface" in spacetime. Events such as D are said to be *simultaneous* with event A.

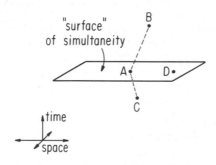

Figure 6   A spacetime diagram illustrating the notion of simultaneity with event A (see text).

The key consequence of assumption (*ii*) is that it gives one an absolute notion of simultaneity, that is, it allows one to give a definition of the phrase "at the same time" without making reference to the observation of a particular observer. There can be no disagreement between observers over where the "surfaces of simultaneity" are in spacetime. (Compare this result with chap. 2.) In this manner, spacetime naturally breaks up into space and time.

1.6   Additional Assumptions

To proceed further, we must be more quantitative and employ clocks and meter sticks in our investigation of spacetime. Let O_1 and O_2 be inertial observers. As illustrated in figure 7, O_1 can measure the velocity of O_2 by measuring O_2's spatial position with meter sticks at

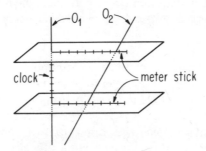

Figure 7   A spacetime diagram illustrating O_1's determination of the velocity of O_2. On the bottom "surface" of simultaneity, O_1 records the position of O_2 on his meter stick. He then counts off $\triangle t$ seconds on his clock and again measures the position of O_2 at this later time. The velocity of O_2 is given by $v = \triangle x / \triangle t$, where $\triangle x$ is the difference between the two meter-stick readings.

two different times. I wish to point out that even here we are using assumption (*ii*), since we have to say which event on O_2's world line is "at the same time" as the reading on observer O_1's clock.

I shall now state two further key assumptions about spacetime. These assumptions will remain valid in special relativity (chap. 2) but will be dropped when the effects of gravitation are taken into account in general relativity (chap. 3).

Assumption (iii): The spatial relationships between simultaneous events are described by Euclidean geometry.

Assumption (iv): Inertial observers all move with constant (uniform) velocity with respect to each other. Furthermore, any observer who moves with uniform velocity with respect to an inertial observer is himself an inertial observer.

1.7   Conclusions

The spatial interval (that is, distance) between two simultaneous events has a well-defined, observer-independent meaning. However, this is not true for two nonsimultaneous events. An observer who is present at both of these nonsimultaneous events would naturally assign zero spatial interval to the events, since, with respect to his frame of reference, they both occur at the same place; in a similar manner, other observers would naturally assign a nonzero spatial interval to these same events. Only in the case where the events are simultaneous would all observers agree.

However, the assumptions given above imply that all inertial observers must agree on the time interval between any two events. The reason is as follows. The surfaces of simultaneity are absolute. If different inertial observers recorded different time intervals between two surfaces of simultaneity, it would be possible to distinguish between them in a manner which would violate the principle of relativity.

I conclude this chapter by giving an answer to our question, What are the intrinsic, observer-independent, properties of spacetime? Based on the assumptions we have made here, the time interval between two events is observer-independent. For simultaneous events, the spatial interval also has intrinsic, observer-independent meaning. These are the only observer-independent characterizations of the spacetime relationships between events.

Two

Special Relativity

Just as there is nothing wrong with Euclid's mathematics, there is nothing logically inconsistent with the assumptions and conclusions about spacetime which we stated in chapter 1. However, the physical world does not conform to these assumptions. Assumption (*ii*) of chapter 1 is simply false: There *is* a limit to how fast an object can move. The consequences of this fact form the subject matter of this chapter.

2.1 Special Relativity

As many experiments and observations have shown, the set of events which can be reached from or which can reach a given event A does not have the structure shown in figure 6. Rather, the structure is that shown in figure 8. It is still possible for an observer or a material body to go from event A to events such as B, or to get to A from events such as C. However, the set of events which cannot be connected to A in this manner forms more than a three-dimensional surface in spacetime. All events, such as D and E, which lie outside the "cone" drawn in figure 8 (called the *light cone* of A for reasons which will be made clear shortly) cannot be connected to A. Events such as D and E are said to be *spacelike* related to A. Events which lie interior to the light cone, such as B and C, are said to be *timelike* related to A. Events which lie on the light cone are said to be *lightlike* or *null* related to A.

The key consequence of the above fact is that *there is no absolute notion of simultaneity*. The light cones replace the surfaces of simultaneity as absolute surfaces in spacetime. The absolute notions of spacelike, timelike, or lightlike related replace the absolute notion of "at the same time."

The basic meaning of figure 8 is, of course, that there is a finite, maximum possible velocity which an observer or material body can

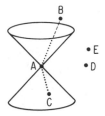

Figure 8   A spacetime diagram showing the spacetime relationship between event A and other events (see text).

attain. However, before we can make this statement precise, we must reexamine the notion of velocity, since with the new notions of the structure of spacetime, even the definition of velocity is not completely straightforward. With our old notions, inertial observer O_1 determined the velocity of inertial observer O_2 as follows (see fig. 7): O_1 recorded the position of x of O_2 at some initial time t and O_2's position x' at a later time t', and he defined O_2's velocity as $v = (x' - x)/(t' - t)$. In precise terms, by "the position of O_2 at time t," we meant the position of the event on O_2's world line which lies on the same absolute surface of simultaneity as the event on O_1's world line at which his clock reads t. But now we have learned that there are no such absolute surfaces of simultaneity in spacetime. In order to define velocity we must give a prescription for determining which event on O_2's world line corresponds to a reading on O_1's clock.

Fortunately, there is a natural—though rather complicated—prescription for this. First we must explicitly assume that any observer O_{1a} who maintains a constant spatial position with respect to inertial observer O_1 is himself an inertial observer. We can synchronize the clocks carried by O_1 and O_{1a} by any method which treats O_1 and O_{1a} in a symmetrical manner. (For example, as illustrated in fig. 9, we can station yet another inertial observer halfway between O_1 and O_{1a} and have him send a signal in a symmetrical manner to both O_1 and O_{1a}. O_1 and O_{1a} are instructed to set their clocks to $t = 0$ when they receive this signal.) In this manner, O_1 can set up a whole family of other inertial observers, all of whom maintain a constant spatial position with respect to him and have

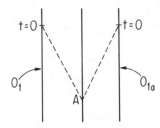

Figure 9   A spacetime diagram illustrating a method by which observers O_1 and O_{1a} can synchronize their clocks. At event A, an observer halfway between O_1 and O_{1a} sends a signal to them and they each set their clock to zero when they receive it.

their clocks synchronized with his. Now observer O_1 can label all events in spacetime by the following prescription: Given an event E, there will be an inertial observer O_{1b} in this family whose world line passes through E. Label event E by the spatial position of O_{1b} and the reading on the (synchronized) clock of O_{1b} at E. This prescription for the labeling of events is referred to as the *global inertial coordinate system* associated with inertial observer O_1. This global inertial coordinate system gives one a notion of simultaneity: two events are considered simultaneous if they are labeled by the same time value t in O_1's global inertial coordinate system. Given this notion of simultaneity, O_1 can define the velocity of O_2 in the same manner as before.

Now that we have a notion of simultaneity, it may seem that we are back to the situation described in chapter 1. However, this is not true. The prerelativity notion of simultaneity was *absolute;* here the notion of simultaneity is *relative,* that is, it depends on a construction performed by a particular inertial observer. As we shall see in more detail in the next section, inertial observers who move at nonzero velocities with respect to each other will disagree over which events are "at the same time" as a given event.

With this framework established, we may now state the new assumption (ii') which replaces the prerelativity assumption (ii) of chapter 1.

Assumption (ii'): No object can have a velocity greater than a certain maximum possible velocity c. In vacuum, light always travels with this velocity. (The numerical value of this fundamental constant of nature c, the speed of light, is 186,000 miles per second.)

The reason for the terminology *light cone* used above should now be apparent. Events which lie on the light cone of event A are precisely those which can (in principle) be connected to A by a light signal.

Note that assumption (*ii'*) states that *all* inertial observers must obtain the value c when they measure the speed of light. It would violate the principle of relativity (assumption (*i*) of 1.4) to postulate that there is a preferred inertial observer who always measures c, while other inertial observers do not. Historically, it was believed for a time that there was such a preferred inertial observer who was described as "at rest with respect to the ether." However, in 1887 Michelson and Morley performed their famous experiment designed to determine with high precision the change in the measured velocity of light due to motion of an observer through the ether. They found no change whatsoever. As a result of this experiment, the ether theory was overthrown and the principle of relativity was confirmed.

Notice also that assumption (*ii'*) implies that the path of a light signal in spacetime must be independent of the motion of the emitter of the signal; otherwise the velocity of light would depend on the motion of the emitter and would not always be c.

How is it possible that two inertial observers who are moving with nonzero velocity with respect to each other find that the same light ray moves past each of them with the same velocity c? Admittedly, it takes a while to get used to this fact, but there is nothing paradoxical about it. Recall that the definition of velocity given above crucially involves the notion of simultaneity. In special relativity, simultaneity is not absolute; different inertial observers have different notions of simultaneity. Hence, the simple law of addition of velocities, which holds in prerelativity physics by virtue of the absolute notion of simultaneity, does not hold in special relativity. In their measurement of the speed of light, the difference in velocity of the two inertial observers is exactly compensated by their disagreement over simultaneity, and they both obtain the result c. (Frequently, facts such as this are explained in terms of "shrinking meter sticks"

[Lorentz contraction] and "slow-running clocks." However, these explanations are quite confusing and misleading. These effects will be discussed in 2.5.)

To complete the specification of the key postulates of special relativity, we carry over assumptions (*iii*) and (*iv*) in essentially unchanged form.

Assumption (iii): The spatial relations between (relatively) simultaneous events, that is, events which are determined to be simultaneous by any inertial observer using the procedure described above, are described by Euclidean geometry.

Assumption (iv): Inertial observers all move with constant (uniform) velocity with respect to each other. Furthermore, any observer who moves with uniform velocity with respect to an inertial observer is himself an inertial observer.

Assumptions (*i*) (see 1.4), (*ii'*), (*iii*), and (*iv*) are the essential content of the special theory of relativity. In the remainder of this chapter, we shall elaborate on some of the consequences of these assumptions and state the key results in a useful form for our discussion of the theory of general relativity in chapter 3.

2.2   Relative Simultaneity

In this section we shall illustrate in greater detail the fact that different inertial observers disagree as to which events occur simultaneously with a given event.

We can continue to investigate the properties of spacetime by using clocks, meter sticks, and families of inertial observers who maintain a constant spatial position with respect to a given inertial observer O. However, assumption (*ii'*) allows us to get the same information in a much more straightforward manner by using only clocks and light rays as follows.

Consider two events, A and B, and an inertial observer O who passes through event A (see fig. 10). If a light signal were emitted at event B, observer O would see this signal at event D. Similarly, there will be an event C on O's world line having the property that a light signal emitted at C will reach B. Let t_1 be the time interval—as determined by observer O's clock—between events A and D; let t_2 be

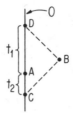

Figure 10   A spacetime diagram illustrating how observer O can determine space and time intervals between events A and B by using only clocks and light signals (see text).

the time interval between C and A (t_1 or t_2—but not both—can be negative or zero, depending on the spacetime relationship of events A and B; see fig. 12). Let

$$\Delta t = (t_1 - t_2)/2$$
$$\Delta x = c(t_1 + t_2)/2$$

Then Δt and Δx, respectively, are equal to the time interval and space interval between events A and B as determined by observer O, that is, they agree with the expressions O would have obtained if he had used meter sticks and a family of inertial observers with synchronized clocks as described in 2.1. In practice, we determine the distance to the moon in this manner: we bounce a radar or laser signal off the surface of the moon and use the above formula for Δx.

Events which are simultaneous with event A as determined by observer O are those for which $\Delta t = 0$, that is, $t_1 = t_2$. Now we can see why an inertial observer O and another inertial observer O', who moves with nonzero velocity with respect to O, disagree as to which events are simultaneous with event A.

Figure 11a shows the events which are simultaneous with A as determined by O. In particular, O finds that E is simultaneous with A ($\Delta t = 0$), while F occurs later ($\Delta t > 0$). In figure 11b we consider the same events and examine the relationship between them as determined by O'. As I remarked in 2.1, the path of a light ray in spacetime is independent of the motion of the emitter. Thus in figure 11b the path of the light rays which are sent to and from events E and F are the same as in figure 11a. But simultaneous events as

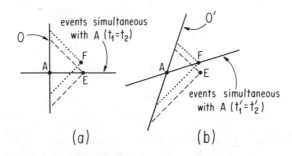

events simultaneous
with A ($t_1 = t_2$)

events simultaneous
with A ($t'_1 = t'_2$)

(a)   (b)

Figure 11   Spacetime diagrams showing that observers O and O′ disagree over which events are simultaneous with A (see text).

determined by O′ are those for which $t'_1 = t'_2$. Hence, we can see from figure 11b that O′ finds that A and F are simultaneous ($\Delta t' = 0$) while event E occurred earlier ($\Delta t' < 0$)!

Incidentally, I have drawn figure 11 taking the point of view that O is "at rest" and O′ is "moving." Of course, by the principle of relativity there is no absolute way of determining "at rest." It is my prejudice in making the drawing, not the physics of spacetime, which makes figure 11 look unsymmetrical for observers O and O′.

2.3   The Spacetime Interval

We now return to the basic question posed in chapter 1: What are the intrinsic, observer-independent properties of spacetime?

Previously, under the assumptions of chapter 1, we found that the time interval between two events is observer-independent. Also, for simultaneous events the space interval was found to be observer-independent. However, we have already seen that with our new assumptions, the time interval between two events, $\Delta t = (t_1 - t_2)/2$, is not observer-independent. As described in 2.2, if inertial observers O and O′ have nonzero relative velocity, then in general $\Delta t \neq \Delta t'$, that is, they disagree on the time interval. What quantity or quantities are observer-independent in special relativity?

Key Result: The above assumptions (i), (ii'), (iii), and (iv) imply that the quantity $t_1 t_2$ is observer-independent, where t_1 and t_2 are defined by the construction given in figure 10. In other words, for any two inertial observers O and O′ we have $t_1 t_2 = t'_1 t'_2$.

Unfortunately, I do not know of any simple argument which shows why this result follows from the assumptions. However, for the benefit of the reader who has had some previous acquaintance with special relativity, I mention that one can prove this result by showing that the global inertial coordinate systems defined by any two inertial observers must be related to each other by a Lorentz transformation.

The meaning of the quantity $t_1 t_2$ can be made clearer by noting that

$$t_1 t_2 = \left(\frac{t_1 + t_2}{2}\right)^2 - \left(\frac{t_1 - t_2}{2}\right)^2 = \left(\frac{\Delta x}{c}\right)^2 - (\Delta t)^2$$

so that $t_1 t_2$ is the difference between the squares of the space interval Δx (divided by c) and the time interval Δt. We call $t_1 t_2$ the *spacetime interval* between the two events.

What immediate information does the spacetime interval give one? If the spacetime interval between events A and B is negative, then either t_1 or t_2 is negative. It follows that events A and B are timelike related, as illustrated in figure 12a. In this case it is possible for an inertial observer to be present at both events A and B. The elapsed time such an observer would measure between A and B is

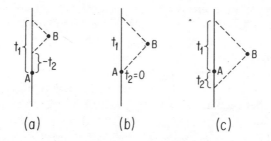

(a)   (b)   (c)

Figure 12   Spacetime diagrams indicating the various possibilities for the sign of the spacetime interval between two events (see text).

simply the square root of minus the spacetime interval, $\Delta t = \sqrt{-\text{(interval)}}$.

If the spacetime interval between A and B is zero, then either $t_1 = 0$ or $t_2 = 0$. If $t_2 = 0$, as in figure 12b, it is possible to send a light signal from A to B; if $t_1 = 0$, it is possible to send a light signal from B to A.

If the spacetime interval is positive, then A and B are spacelike related (fig. 12c). In this case there exists an inertial observer who determines these events to be simultaneous. The distance between the events as determined by this observer is simply the square root of the spacetime interval times c, $\Delta x = c\sqrt{\text{(interval)}}$.

2.4   The Geometry of Spacetime

In this section, we shall restate some of the above results in a slightly different language. Although this restatement does not introduce any new content to the theory, it does carry with it a new point of view on spacetime. When this point of view is developed further (see chap. 3), it leads to a new revolution in the theory of space and time which is every bit as great as the revolution introduced by assumption (*ii'*).

The new language is the following. Call the formula for the spacetime interval between two nearby events the *metric of spacetime*. The metric of spacetime is completely analogous in its mathematical properties to the metric of an ordinary geometrical surface such as a sphere (1.3), with one major exception: Whereas the metric of an ordinary surface is positive since (distance)² is always positive, the metric of spacetime is not always positive since the spacetime interval between timelike related events is negative, and the spacetime interval between lightlike related events is zero.

It is not difficult to write out the spacetime metric explicitly. Label the events in spacetime by the global inertial coordinates t, x, y, and z associated with any inertial observer O. Then, we have

$$ds^2 = -dt^2 + \frac{1}{c^2}(dx^2 + dy^2 + dz^2).$$

Here, ds^2 is the spacetime interval of the nearby events, dt is the time interval between the events as determined by O, and dx, dy, and dz are the three spatial intervals. The above formula is nothing more than our previous formula for the spacetime interval in the case of nearby events.

We can use this new language to give our final answer to the basic question, What are the intrinsic, observer-independent properties of spacetime? *The intrinsic properties of spacetime are fully described by "the geometry of spacetime" as embodied in the spacetime metric.*

2.5   Lorentz Contraction and Time Dilation

One of the most widely discussed effects in special relativity is Lorentz contraction, the "shrinking of meter sticks of moving observers." In this section, we shall explain what is meant by this.

Consider two inertial observers, O_1 and O_2, who move with nonzero relative velocity. Suppose O_2 carries a meter stick with him. A spacetime diagram showing the world lines of O_1, O_2, and the ends of O_2's meter stick is given in figure 13.

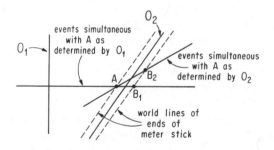

Figure 13   A spacetime diagram illustrating the phenomenon of Lorentz contraction.

What is the length of O_2's meter stick? Before trying to answer this question, we should restate it with more precision, What is the spatial distance between the event representing one end of the meter stick at a certain time and the other end of the meter stick *at the same time?* In prerelativity physics, all observers agree on the notion of simultaneity and on the spatial distance between simultaneous events. But in special relativity, O_1 and O_2 disagree as to which event on the opposite end of the meter stick is "at the same time" as event A in figure 13. Observer O_1 says that B_1 is simultaneous with A, while O_2 says B_2 is simultaneous with A.

Thus, observer O_1 says that the length of O_2's meter stick is c times the square root of the spacetime interval between A and B_1. But O_2 says that the length of his meter stick is c times the square root of the spacetime interval between A and B_2. From our formula for the spacetime interval, it is not difficult to show that the length determined by O_1 is smaller than the length determined by O_2. Thus O_1 determines that the length of O_2's meter stick is less than 1 meter! Hence, O_1 might say that O_2's meter stick has "shrunk as a consequence of O_2's motion." Of course, it is quite misleading for O_1 to say this. Nothing has "shrunk"; all that has happened is that O_1 and O_2 disagree over simultaneity. This is the phenomenon of Lorentz contraction.

There is a similar effect called time dilation or "the slowing down of clocks of moving observers." If O_1 and O_2 carry clocks, O_1 can ask about the change in the reading of O_2's clock during a given time interval Δt on O_1's clock. Again, a precise formulation of this question requires a notion of simultaneity. With O_1's definition of simultaneity, the change in the reading of O_2's clock will be less than Δt, that is, O_2's clock "runs slower." On the other hand, with O_2's definition of simultaneity, O_1's clock "runs slower" (just as with O_2's definition of simultaneity, a meter stick carried by O_1 will have "length" less than 1 meter).

As should be clear from the above discussion, the phenomena of Lorentz contraction and time dilation are not paradoxical. They are merely reflections of the fact that there is no absolute notion of simultaneity in special relativity, and hence different observers will give different answers to questions involving the notion of simultaneity. Lorentz contraction and time dilation may seem anti-intuitive at first because relative simultaneity is anti-intuitive until one gets used to this notion.

The best advice one can give concerning how to understand effects which occur in special relativity is the following: Whenever possible, formulate all statements in terms of the absolute, intrinsic structure of spacetime, namely the geometry of spacetime as embodied in the spacetime metric. If this is done, all statements will be observer-independent and no paradoxes arise. On the other hand, if one makes implicit use of nonabsolute notions such as simultaneity in the formulation of statements, then one is likely to get into trouble. The next section gives a good illustration of a traditional "paradox" of special relativity which can be analyzed in a straightforward, unambiguous manner when the "spacetime geometry" point of view is taken.

2.6   The Twin "Paradox"

Consider two observers O and O' who are both present at events A and B in spacetime, but take different paths in spacetime to get from A to B (see fig. 14). Usually O is taken to be an inertial observer and O' is taken to be his twin who flies off in a rocket ship (noninertial motion) at event A and rejoins his twin at event B. Suppose O and O' have identical clocks and synchronize them at event A. How will the readings of the two clocks compare when they rejoin at event B?

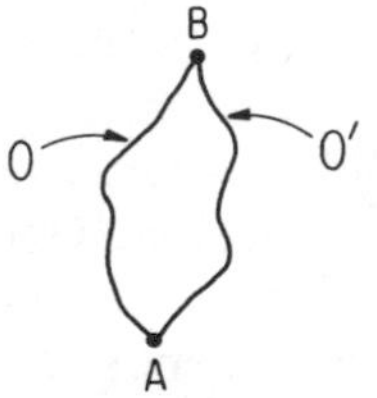

Figure 14   A spacetime diagram depicting the world lines of the twins O and O' between events A and B.

In prerelativity physics, the two clocks will, of course, read the same when they rejoin. But in special relativity, this is not true; in general, the readings of the clocks will differ at event B.

How does one analyze this effect? Let us go back to 1.3, the geometry of a sphere, for a moment and answer an analogous question. Suppose we have two paths between two points on a sphere, for example, two routes between Moscow and New York. Which path is shorter? We can answer this question as follows: Break up each path into a sequence of short, nearly straight segments. Use the metric of the sphere to find the length of each segment, and add up the lengths of all the segments to get the total length of the path. The path with the smaller total length is the shorter path.

We can analyze the twin "paradox" in exactly the same manner. Break up the path in spacetime of each twin into a sequence of small, nearly straight segments. Now use the metric of spacetime to calculate the elapsed time for each segment, and add up these values to get the total elapsed time associated with each path. Just as different paths between two points on a sphere have different lengths, different motions between two events in spacetime will have different total elapsed time. So, in general, one of the twins will be younger than the other twin when they rejoin at event B!

2.7 Geodesics in Spacetime

In 1.3 I mentioned that geodesics on the sphere are defined as the lines which curve as little as possible, that is, the straightest possible lines. Furthermore, one can show that the shortest path between two points on a sphere is a geodesic. I also have already noted in 2.4 that the mathematical structure of the metric of spacetime is closely analogous to the metric of ordinary surfaces, except that the spacetime metric is not positive. It turns out that this difference does not affect the definition of geodesics. One can define geodesics in spacetime as the lines in spacetime which curve as little as possible. What are the properties of these geodesics in spacetime?

Geodesics in spacetime are classified as spacelike, null (lightlike), or timelike according to whether the spacetime relationship between two nearby events on the geodesic is spacelike, null (lightlike), or timelike. The physical significance of each of these types of geodesics is as follows: If two inertial observers who are at rest with respect to each other stretch a rope between them, the configuration of the rope on a surface of simultaneity (as determined by these inertial observers) will be a spacelike geodesic. Null geodesics and timelike geodesics have a much more direct significance. *The path of a light ray in spacetime is described by a null geodesic. Inertial observers are precisely the observers who move on timelike geodesics of the spacetime metric.* In other words, the "straightest possible" motion of an observer in spacetime is nonaccelerating (inertial) motion. The fact that motions of inertial observers are identified with timelike geodesics of the spacetime metric plays a very important role in the formulation of general relativity.

Just as the *shortest* path between two points on a sphere is a geodesic, in spacetime the path with the *longest* elapsed time between two timelike related events is a timelike geodesic. Thus, in the twin "paradox," an observer undergoing accelerated motion always ages less than an inertial observer.

In Euclidean geometry, initially parallel geodesics always remain parallel, that is, parallel straight lines never meet or diverge from each other. One describes this situation by saying that Euclidean geometry is *flat*. On the other hand, initially parallel geodesics on the sphere begin converging toward each other and eventually cross. One says, therefore, that the geometry of the sphere is *curved*. By this criterion, is the geometry of spacetime flat or curved?

As we stated above, timelike geodesics in spacetime are simply the paths of inertial observers. Two initially parallel timelike geodesics correspond to two inertial observers who are initially at rest (zero velocity) with respect to each other. In special relativity inertial observers who initially are at rest always remain at rest with respect to each other, that is, they maintain a constant spatial separation. Thus, initially parallel timelike geodesics always remain parallel. It is not difficult to show that the same must be true of null and spacelike geodesics. *Thus, in special relativity the geometry of spacetime is flat.*

Three

General Relativity

The theory of special relativity, developed by Einstein in 1905, gave rise to a truly major revolution in our notions of space and time. The effects of this revolution on the known laws of physics were soon felt, as theorists began to reformulate them within a framework compatible with special relativity. The attempt to reformulate the theory of gravitation produced a second major revolution in our notions of space and time: Einstein's theory of general relativity, developed in 1915.

3.1   Special Relativity and Physical Theories

As I previously remarked, the notion of absolute simultaneity is so deeply ingrained in the way most people think about spacetime that it even takes a great deal of effort to be consciously aware of when and how one is using this assumption. Hence, it is not surprising that the known laws of physics prior to special relativity implicitly assumed absolute simultaneity in their formulation. For example, Newton's law for the gravitational force between two bodies states that

$$F = -G \frac{m_1 m_2}{r^2}.$$

Here F is the force, G is the gravitational constant, m_1 and m_2 are the masses of the two bodies, and r is the distance between them. More precisely, F is supposed to be the force between the bodies at a given time and r is supposed to be the distance between the bodies *at this same time*. Within the framework of the prerelativity notions of space and time, this is a perfectly meaningful statement. But, in special relativity, the phrase "at the same time" has no absolute meaning. Newton's theory of gravity does not make sense because it is phrased in terms of spacetime structure which does not exist!

Clearly Newton's theory of gravity, as well as most theories of physics prior to special relativity, can at best be only approximations —valid when all relevant velocities are small compared with c—to the true exact laws of physics. The aim now is to find these true laws of physics.

What are the properties of the new laws of physics which we are seeking? First, they must be formulated within the framework of the spacetime structure of special relativity. If notions such as relative simultaneity are used in the formulation, it must be demonstrated that the theory is actually independent of which inertial observer is chosen to define simultaneity. (Otherwise, the principle of relativity would be violated.) This is usually referred to as "relativistic invariance" or "covariance" of the theory. Second, the predictions of the theory must be consistent with special relativity. In particular, the theory must not predict that one can send a body or signal between two events which are spacelike separated.

The theory of electricity and magnetism given by Maxwell over a century ago satisfies these requirements. Indeed, historically the analysis of Maxwell's theory and its predictions concerning the propagation of electromagnetic waves (that is, light) led to the discovery of special relativity. What is the situation for the theory of gravitation?

The close analogy between Newton's theory of gravitation and the theory of electrostatics is well known. Coulomb's law for the force between two charges differs from Newton's force law only by a change in sign. Hence, one might hope that one could formulate a special relativistic theory of gravitation closely analogous to Maxwell's theory of electromagnetism. In the limit where the motion of the masses is slow compared with the speed of light, this theory should reduce to Newton's theory, just as in the slow-motion limit Maxwell's theory reduces to Coulomb's electrostatics. However, this attempt fails. It turns out that the sign difference between gravity and electromagnetism also enters the formula for the energy of radiation. Gravitational radiation in a theory of gravity analogous to Maxwell's theory would carry away negative energy. This is physically unacceptable, since it means that a system of masses that radiates would *gain* energy, which would cause it to radiate more and gain more energy ad infinitum.

One might not become terribly discouraged by this and try to formulate a theory of gravitation in the framework of special relativity which does not suffer from the above difficulty. In fact, it is possible to do so, though other difficulties then arise. However, there are a number of reasons for believing that perhaps one should not simply try to fit gravity into the framework of special relativity. These reasons motivated Einstein to abandon some of the assumptions of special relativity and formulate a new theory of space, time, and gravity: general relativity.

3.2   Space, Time, and Gravity

As I remarked above, in Newtonian theory gravity is treated as a field which is closely analogous in many of its properties to an electric field. However, there are two basic ideas which suggest that perhaps one should think of gravitation in a different manner from other fields.

First, gravity is universal; *all* bodies are affected by gravity. (On the other hand, electric fields, for example, do not affect neutral bodies.) Furthermore, all objects fall the same way in a gravitational field. This fact, known as the *equivalence principle,* is expressed in Newtonian theory by the statement that the gravitational force on a body is proportional to its mass. The equivalence principle was first demonstrated by Galileo's famous experiment of dropping two bodies of different weight from the Leaning Tower of Pisa. Much more recently it has been verified to very high precision by experiments of Robert Dicke and others. What is the significance of this universality of gravity? It suggests the possibility of ascribing the properties of the gravitational field to spacetime itself!

A second rather vague set of ideas deals with inertia and goes under the name of *Mach's principle,* though many of the notions predate Mach's nineteenth-century writings. This basic idea is as follows: Consider a rotating body. One can measure its rotation in two independent ways: (1) in an absolute, local manner by measuring the stresses in the body required to keep the outer parts of it from flying off, or (2) in relation to distant matter (the "fixed stars"). Why is it that these two notions of rotation agree? Mach's idea was that perhaps the distant matter in the universe determines local inertial effects such as rotation. Thus, if someone were somehow to accelerate the distant matter in the universe, it should affect our local determinations of nonaccelerating and nonrotating. If there were no matter in the rest of the universe, there should be no such thing as inertia or rotation.

Einstein accepted the basic idea of Mach's principle. However, Mach's ideas do not have any expression in special relativity where the structure of spacetime is not affected by the matter which is present. Thus, Einstein was motivated to seek a new theory in which the effects of gravitation would be expressed in terms of the structure of spacetime (equivalence principle) and the structure of spacetime would be determined by the matter present in spacetime (Mach's principle). His success in this regard was probably the greatest single achievement that has yet occurred in theoretical physics.

Interestingly, although some of Mach's ideas are reflected in Einstein's theory of general relativity (see 8.4), it is certainly not true that general relativity fully incorporates all of them. In general relativity, it can be said that the motion of distant matter affects some local inertial properties but it does not fully determine them. In particular, rotation can always be locally determined even if there is no distant matter.

3.3   General Relativity

I now shall describe the theory of general relativity, Einstein's theory of space, time, and gravitation. Recall, first, two basic results of our discussion of special relativity: (1) The intrinsic properties of spacetime are fully described by the spacetime metric (see 2.5). (2) Inertial observers are precisely those observers who move on timelike geodesics of the spacetime metric (see 2.7). Recall also two of the key assumptions which went into the formulation of special relativity (see 2.1): (*iii*) The spatial relationships between (relatively) simultaneous events are described by Euclidean geometry. (*iv*) Inertial observers all move with uniform velocity with respect to each other; furthermore, any observer who moves with uniform velocity with respect to an inertial observer is himself an inertial observer.

Here is Einstein's brilliant idea: Identify freely falling bodies in a gravitational field with the inertial observers of special relativity.

This clearly necessitates dropping assumption (*iv*) since freely falling bodies in a nonuniform gravitational field do not maintain a uniform velocity with respect to each other; drop assumption (*iii*) as well. However, retain the two key conclusions, 1 and 2, stated above.

In other words, we continue to assume that all the observer-independent information concerning the spacetime relationship between events is contained in the metric of spacetime. However, we no longer assume that the spacetime metric is given by the simple expression

$$ds^2 = -dt^2 + \frac{1}{c^2} (dx^2 + dy^2 + dz^2)$$

as it is in special relativity (see 2.4). Such a spacetime metric corresponds to a flat geometry in that initially parallel geodesics remain parallel (see 2.7). In general relativity we wish to retain the identification of the world lines of inertial (that is, freely falling) observers with geodesics of the spacetime metric. However, in a (nonuniform) gravitational field, two freely falling observers who are initially at rest do not remain at rest with respect to each other. Thus, in general relativity the presence of a gravitational field is reflected in the fact that initially parallel geodesics of the spacetime metric do not remain parallel. In other words, *spacetime is curved.*

Note that dual role of the spacetime metric in general relativity: (*a*) Just as in special relativity, it carries all the information concerning the spacetime relationship of events; (*b*) It fully describes the gravitational field by specifying, via its timelike geodesics, the motion of all freely falling observers. Thus, the description of gravitation is inextricably tied into the properties of space and time in that both are described by a single quantity, the spacetime metric. The presence of a gravitational field corresponds to curvature of the spacetime geometry.

3.4   Einstein's Equation

The theoretical framework of general relativity was outlined in the previous section. The gravitational field is to be described in terms of the curved geometry of spacetime. To complete the specification of the theory, it remains to specify what spacetime geometry (that is, gravitational field) is associated with a given configuration of matter.

Einstein did this by postulating an equation which, in essence, says,

"curvature of spacetime" = "energy density of matter."

Thus, Einstein provided us with a truly remarkable and beautiful theory of gravitation where in accordance with the ideas of 3.2, the effects of gravitation are fully expressed in terms of the structure of spacetime, and, in accordance with some of Mach's ideas, the structure of spacetime is related to the distribution of matter via Einstein's equation.

This description of Einstein's equation is somewhat of an over-simplification and some words of caution should be given. First, the left side of the equation is not the entire curvature of spacetime but only a part of it. Thus, outside of the matter distribution (where the right side of Einstein's equation will be zero) spacetime will in general still be curved, that is, a gravitational field will be present. Furthermore gravitational radiation—ripples in the curvature which propagate through spacetime—can exist (see 9.4). Second, the right side contains contributions from other properties of matter besides energy density. In particular, pressures and stresses contribute to the curvature of spacetime in general relativity.

If no matter is present, the right side of Einstein's equation vanishes and a perfectly valid solution is the flat spacetime geometry of special relativity. (However, there also are many other solutions describing possible exterior gravitational fields of bodies, gravitational waves, and so on.) In this sense, general relativity includes special relativity as a special case.

In our description of general relativity it was postulated that freely falling bodies follow geodesics of the spacetime metric. It turns out, however, that—as discovered more than ten years after the formulation of general relativity—Einstein's equation itself actually determines the motion of matter in spacetime. The "geodesic hypothesis" actually follows as a consequence of Einstein's equation and does not have to be postulated separately.

Finally, it should be mentioned that in practice it has proven very difficult to obtain exact solutions of Einstein's equation. Fortunately, many solutions of great physical interest (for example, solutions describing black holes) are known and the properties of these space-times have been investigated in detail. But the study of what possible structures of spacetime are permitted in general relativity continues to be hampered by our general inability to solve Einstein's equation.

3.5   Consequences of Einstein's Theory

Einstein's theory of general relativity is unquestionably one of the most beautiful theories ever devised, but that, of course does not mean that nature conforms to it. In this section we shall briefly describe three famous predictions of the theory which are subject to experimental and observational check. All these predictions have been verified. One would not yet be justified in saying that the validity of general relativity has been proven beyond any doubt, since the theory has not yet been tested for very strong gravitational fields and the accuracy of the experiments and observations which have been done is not great enough to rule out small modifications to Einstein's equation. However, the agreement between theory and observation in these three cases described below greatly increases one's confidence that general relativity is a true theory of nature.

The Motion of the Planets

If the energy density of matter is not too large, it is possible to obtain approximate solutions of Einstein's equation for the spacetime geometry. One can then calculate the geodesics of the spacetime metric to obtain the motion of freely falling bodies. In the limit where the relative velocities of the bodies of matter is much smaller than c, one finds that the predictions of general relativity reduce to the predictions of Newton's theory of gravitation. Thus, in particular, in this approximation general relativity predicts that the planets should move around the sun in elliptical orbits which obey Kepler's laws.

However, when corrections to the above approximations are put back in, one finds that the predictions of general relativity deviate from those of Newtonian theory. In particular, the elliptical orbits of planets don't quite close but precess instead. This precession is much too small to observe for all planets except Mercury, where general relativity predicts an orbital precession of 43 seconds of arc per century. (A second of arc is $\frac{1}{3,600}$ of a degree.) *This precession of Mercury's orbit has been observed.* Indeed, this precession had been observed long before the development of general relativity and had been an unexplained mystery until then.

Light "Bending"

As discussed above, in general relativity, just as in special relativity a freely falling (that is, nonaccelerating) material body moves on a timelike geodesic of the spacetime metric. Similarly, in general relativity it remains true that the path of a light ray in spacetime is null (lightlike) geodesic (see 2.7). Because spacetime is not flat in the vicinity of the sun, a light ray passing near the sun will appear to a distant observer to be deflected (see fig. 15). One can think of this physically as the light being bent by the gravitational attraction of the sun, although, in fact, light is actually traveling on the "straightest possible path" in the curved spacetime geometry.

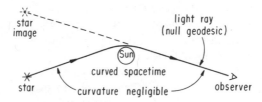

Figure 15   A space diagram (*not* a spacetime diagram) illustrating the "light-bending" effect. Because of the spacetime curvature in the vicinity of the sun, the direction of motion of the light is changed. The deflection angle is greatly exaggerated in this diagram. The actual deflection angle of a light ray passing by the surface of the sun is only 1.75 seconds of arc.

General relativity predicts a deflection angle of 1.75 seconds of arc for light rays which just pass by the surface of the sun. It is difficult to make an accurate observation of this effect. One must wait for a total eclipse of the sun in order to observe the deflected light and a number of observational problems arise. However, the light-bending effect has been observed and found to agree with the predictions of general relativity to within the uncertainty of the observation (about

10% of the predicted value). Recently, the light-bending effect has been investigated for radio waves coming from a quasar as it passes behind the sun. This approach has already verified the light-bending effect to a better accuracy than the eclipse method and it is hoped that in the future it will provide an even more precise check of this prediction of general relativity.

The Gravitational Red Shift

Suppose an observer O_1 sends out two signals with a time interval Δt_1 between them. Let observer O_2 receive these signals. In curved spacetime—or even in flat spacetime if there is relative motion between O_1 and O_2—there is no reason why the time interval Δt_2 between O_2's reception of the signals must equal Δt_1. Thus, in particular, if O_1 emits light at frequency ν_1, O_2 will, in general, observe it to have frequency $\nu_2 \neq \nu_1$. If O_1 and O_2 are "at rest" and $\nu_2 \neq \nu_1$, this effect is known as the *gravitational red shift*. One can show that light emitted in a region of strong gravitational attraction will be seen by a distant observer to have a lower frequency.

Actually, it is not difficult to see that the gravitational red shift must occur if energy is conserved. Quantum theory tells us that the energy, E, of a photon (that is, light) is proportional to its frequency, $E = h\nu$, where h is Planck's constant. If the frequency of light remained unchanged as it left a region of strong gravitational attraction, its energy would not be decreased. One could then convert the energy of the photons in the light beam to rest mass energy and lower the mass back into the strong field region using the gravitational attraction to do work. If one then converts the rest mass energy back into photons, one will return to the same configuration as one began with, but energy will have been gained in the process of lowering the mass. Thus, the quantum formula $E = h\nu$, together with conservation of energy, requires the existence of the gravitational red-shift phenomena predicted by general relativity. The gravitational red-shift effect has been accurately confirmed by experiment.

Four

Implications for Cosmology: The "Big Bang"

For the remainder of this book, we shall assume the validity of Einstein's theory of general relativity. In general relativity, the spacetime geometry—which simultaneously describes both the spacetime relationship between events and the gravitational field—is related to the distribution of matter in the universe via Einstein's equation. The solutions of Einstein's equation represent all possible spacetime geometries. Which solution of Einstein's equation describes our universe? The answer to this fundamental question is the subject of this chapter.

4.1   The Large-Scale Structure of Our Universe

When we look into the sky at night, we see it filled with stars. About one hundred billion stars are clustered in our vicinity[1] to form our galaxy, the Milky Way. The universe is populated with other such vast aggregates of stars. Between the galaxies, there is little, if any, matter. Figure 16 is a photograph of some typical galaxies as seen through a telescope.

Figure 17 shows a spacetime diagram of part of our universe. I have drawn in the world line of our galaxy and the world lines of several other galaxies. What is the basic spacetime structure of our universe?

It is an observational fact that, at least to a rough approximation, over large distance scales the universe appears to be isotropic, that is, it "looks the same" in every direction as viewed from our galaxy. In other words, the nature of what we see when we look at faraway

---

1. By "in our vicinity," I mean within one hundred thousand light years of us. A light year is the distance that light—moving at 186,000 miles per second—travels in one year, so the distance scales considered in this chapter are truly large.

Figure 16   The Hercules cluster of galaxies. Each of the spiral and fuzzy, elliptically shaped objects is a galaxy, containing, on the average, about one hundred billion stars. (Yerkes Observatory photograph; photo taken with Kitt Peak National Observatory four-meter telescope.)

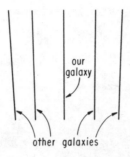

Figure 17   A spacetime diagram of a portion of the universe showing the world lines of our own galaxy and other galaxies.

objects does not depend on what direction in the sky we look. Further strong evidence for isotropy of the universe comes from the fact
that the cosmic microwave background radiation (see 5.1) has
been measured to be isotropic to a very high degree of precision.

In addition, the gross properties of the universe do not appear to
vary with distance from us. We, of course, have no direct way of
knowing the properties of the distant regions of the universe which
are beyond the range of our best telescopes. However, it seems most
natural to suppose that the properties of the universe (for example,
the number of galaxies in a given volume) are roughly the same
everywhere else as they are here. In other words, if we lived on
some faraway galaxy, we would expect the basic observable properties of the universe not to differ greatly from what we see from our
own galaxy. Thus, we shall assume that the universe is roughly
homogeneous. This philosophical belief is referred to as the *Copernican principle,* since Copernicus first taught us that we do not occupy
a special position in the universe.

The assumptions of homogeneity and isotropy of the universe in
the sense described above have an important direct implication.
They imply that there must exist a family of homogeneous and
isotropic three-dimensional spatial "surfaces" in spacetime (see
fig. 18). Each of these surfaces can be thought of as representing
the universe at an instant of time; the collection of all these surfaces
fills all of spacetime. At the instant of time represented by any one

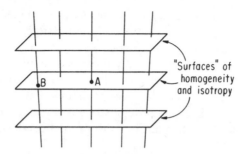

Figure 18   The same spacetime diagram as figure 17 but with the surfaces of homogeneity and isotropy drawn in. By our assumption of homogeneity, the basic observable properties of the universe viewed by an observer in our galaxy at event A are the same as would be seen by an observer in a distant galaxy at event B.

of these surfaces, the nature of what an observer would see does not depend on where in the surface the observer is located. However, our assumptions leave open the possibility that the universe may be changing with time.

Note that the galaxies give us "preferred inertial observers" and the above surfaces give us a preferred notion of simultaneity. How does this square with the discussion of the first three chapters above, where it was concluded that—at least in special relativity—there are no preferred inertial observers and no absolute surfaces of simultaneity? The answer is that one must keep in mind the distinction between a theory and specific solutions admitted by the theory. In general relativity as in special relativity, there are no preferred inertial observers in the theory. However, a particular gravitational field (that is, spacetime geometry) may, by its symmetries or other properties, allow us to single out a preferred family of observers. Similarly, the symmetries of a specific solution, like the solution describing our universe, may allow us to pick out preferred surfaces of simultaneity. Of course, these preferred surfaces are not absolute surfaces in the sense of chapter 1; the communication property of spacetime is still as illustrated in figure 8, not figure 6.

In the case of our universe the geometry of each of these preferred surfaces is such that every point "looks like" every other point and all directions look the same. We now can ask the following mathematical question: What are all the possible three-dimensional spatial surfaces with homogeneous and isotropic geometries? The answer is that there are precisely three possibilities.

1. Ordinary three-dimensional Euclidean flat space. This is the geometry which space was always assumed to possess prior to relativity theory. The curvature of space is zero, and Euclid's postulates are valid. In particular, initially parallel geodesics in flat space remain parallel.

2. The three-dimensional sphere. This is the natural mathematical generalization to three dimensions of the one-dimensional sphere (that is, the circle) and the ordinary two-dimensional sphere (for example, the surface of the earth). The circle is defined as the set of points in a two-dimensional Euclidean plane which lie at a unit distance from the origin, $x_1^2 + x_2^2 = 1$. The two-dimensional sphere is defined as the set of points in three-dimensional Euclidean flat space which lie at a unit distance from the origin, $x_1^2 + x_2^2 + x_3^2 = 1$. Similarly, the three-dimensional sphere is defined mathematically as the set of points in a four-dimensional (!) Euclidean flat space which lie at a unit distance from the origin, $x_1^2 + x_2^2 + x_3^2 + x_4^2 = 1$. The curvature of the sphere (of any dimension) is constant and positive, that is, initially parallel geodesics converge.

3. The three-dimensional hyperboloid. This is also the natural mathematical generalization to a higher dimension of the ordinary one-dimensional hyperbola and two-dimensional hyperboloid. It can be defined as the set of points in a four-dimensional Euclidean space satisfying $x_1^2 - x_2^2 - x_3^2 - x_4^2 = 1$. The curvature of the hyperboloid is constant and negative, that is, initially parallel geodesics diverge.

Note that there is a very important difference in character between possibilities 1 and 3 and possibility 2. The volume of Euclidean space and of the hyperboloid are both infinite; both types of space extend without limit in every direction. On the other hand, the volume of the sphere is finite. The three-dimensional sphere, like the two-dimensional sphere, "closes in on itself." In fact, on the sphere if one goes off on a straight path (geodesic) in any direction, one eventually returns to the starting point. If the universe contains three-dimensional spatial surfaces with the character of possibilities 1 and 3, it is said to be *open;* if its spatial surfaces have the character of possibility 2, it is said to be *closed.*

How can we directly determine whether space is curved? One way would be as follows. In principle (but certainly not in practice) we could stretch a rope between three galaxies so as to form a triangle. We could measure the angles of the triangle. If space is flat, the sum of the angles of the triangle will be exactly 180°, as Euclid taught us. But if space is positively curved (sphere), the sum of the angles will be greater than 180°; if space is negatively curved (hyperboloid) the sum will be less than 180°. Obviously, this is not a very practical method and, in any case, for nearby galaxies the expected deviation from 180° would be extremely small. However, as we shall see in the next section, the curvature of space has an important influence on the dynamics of the universe. Observation of the dynamics of the universe gives information on its spatial structure.

What is the spatial geometry of our universe: Euclidean space, the sphere, or the hyperboloid? At present we cannot give a definitive answer to this question. The idea of a closed universe—finite yet without a boundary—is very aesthetically appealing. On these aesthetic and philosophical grounds most theorists have tended to believe that the spatial geometry of the universe is that of the three-dimensional sphere. However, the best observational evidence that is presently available favors an open universe with hyperboloid geometry (see 5.2).

4.2   The Dynamics of the Universe

From our assumptions of homogeneity and isotropy of the universe, we have deduced that the spatial structure of the universe must either be that of Euclidean three-space, the three-dimensional sphere, or the three-dimensional hyperboloid. To complete the specification of the spacetime geometry of the universe, we must determine the (possibly time-dependent) scale factor a which sets the distance scale in each of the spatial "surfaces" of homogeneity and isotropy. In the case of the sphere and hyperboloid, a also determines the magnitude of the spatial curvature.

In each spatial surface of homogeneity the distance between any pair of galaxies is proportional to a. If a were to take the same value on each surface, the distances between the galaxies would not change with time. On the other hand, if a increased with time, the distances between the galaxies would increase and galaxies would recede from each other with a velocity proportional to the distance between them; if a decreased with time, the galaxies would move toward each other in a similar manner. The motion of the galaxies in these cases can be attributed to expansion or contraction of space itself.

We have not yet used Einstein's equation. The curvature of space-time depends only on a and the nature of the surfaces of homogeneity and isotropy. Hence Einstein's equation relates these quantities to the density ρ and pressure P of matter which is present in the universe.

To write the equations obtained from Einstein's equation explicitly, we label each of the spatial surfaces of homogeneity and isotropy in the universe by the time t measured by a clock in our (or any other) galaxy, that is, we label each surface by the reading on our clock when the world line of our galaxy passes through the surface (see fig. 18). Einstein's equation implies the following relations:

$$- 3 \, \frac{d^2a}{dt^2} = 4\pi G \left(\rho + 3\frac{P}{c^2} \right) a$$

$$3 \left(\frac{da}{dt} \right)^2 = 8\pi G \rho a^2 - 3kc^2$$

We have used the notation of calculus in writing these equations: da/dt is the rate of change of the scale factor a with t, that is, it is the expansion (or contraction) rate of the spatial sections; d^2a/dt^2 denotes the rate of change of da/dt, that is, it is the acceleration of the expansion; G and c denote, respectively, the gravitational constant and the speed of light. Finally, k is a number which takes the value 0 for Euclidean three-space, $+ 1$ for the three-sphere, and $- 1$ for the three-dimensional hyperboloid. In 4.4 we will rewrite these equations in terms of observable parameters.

Here is the great surprise. It would have been natural to assume, in addition to the assumptions of 4.1, that the universe is ever constant in time, that is, that the structure of the universe was always the same as it is now and will always remain the same. Einstein's equation says that this is impossible! If matter is present in the universe, the right side of the first equation must be positive, since the density of matter is always positive and the pressure cannot be negative. Hence, the acceleration of expansion cannot be zero (and, in fact, must be negative). This means that—with the possible exception of a single instant of time when expansion stops and changes to contraction—*the universe must always be expanding or contracting*. The notion of an ever-constant, static universe is incompatible with general relativity; the universe must be dynamic.

Einstein was so unhappy with this conclusion that he modified his equation by adding an extra term, called the cosmological constant. With this additional term, the equations admitted a static universe solution. However, in 1929 Hubble announced the discovery that the galaxies we observe are receding from us, with velocity of recession proportional to the distance of the galaxy from us. This is exactly what we should observe if the universe is expanding uniformly by increasing the overall scale factor a of the spatial sections. Thus, the universe *is* dynamic, as predicted by the original form of Einstein's equation. Einstein's reason for modifying his equation is no longer relevant; in fact, after Hubble's discovery, Einstein considered the introduction of the cosmological constant to be his greatest mistake. Throughout this book, we shall assume the validity of the original form of Einstein's equation, without the cosmological constant.

4.3   The "Big Bang"

As explained above, the universe is observed to be expanding. Thus, in the past the scale factor a was smaller and the density of matter in the universe was greater. We can deduce the behavior of the universe in the past from Einstein's equation, that is, the equations written out explicitly in the previous section. When we do so, we find that if the universe is presently expanding as we observe it to be, then a *finite* time ago the scale factor a must have been zero and the density of the universe must have been infinite. Thus, the universe started with a *"big bang"* a finite time ago! The best observational evidence indicates that the "big bang" took place between 8 and 18 billion years ago (see 5.2).

If the spatial surfaces are three-dimensional spheres (closed universe), the volume of the universe goes to zero at the time of the big bang. In the case of flat or hyperboloid geometries (open universe) the volume of the universe is always infinite. However, in all cases the matter density and the curvature of spacetime become infinite at the time of the big bang. Such an exotic situation is referred to as a *spacetime singularity*. We shall discuss spacetime singularities further in 4.5 but remark here that the observational evidence (see chap. 5) leaves little, if any, doubt that the universe has indeed expanded from a much more condensed and dense state.

What happened before the big bang? This is a natural question to ask within the context of pre-general relativistic notions of space and time. But within the context of general relativity, this question does not make sense since there was no such thing as space and time "before" the big bang. The nonsingular points in a solution of Einstein's equation represent *all* events that occur in the universe so it does not make sense to ask about other events. Thus, literally speaking, nothing happened before the big bang. Note that it is difficult even to express this idea, since our language is attuned to prerelativistic notions of space and time. In particular, the word *before* presupposes the notion of time.

If the spatial surfaces are spheres ("closed universe"), then $k = +1$ in the equations of 4.2. Analysis of the equations shows that the expansion of the universe will eventually come to a halt and switch over to contraction. The contraction of the universe then will continue at an ever-increasing rate until, within a finite time, one gets another infinite density, infinite curvature spacetime singularity. Incidentally, this contraction to a singularity occurs just rapidly enough to prevent the possibility of traveling all the way around the universe. In the case of an open universe (flat, $k = 0$; or hyperboloid, $k = -1$) the expansion of the universe continues forever. In the case of flat spatial geometry, the expansion rate asymptotically approaches zero at late times, while in the case of hyperbolic geometry the expansion rate asymptotically approaches a finite, nonzero value. Thus, the nature of the spatial surfaces has a very important effect on the dynamics of the universe.

4.4   Hubble's Law; The Parameters *H* and *q*

As mentioned in 4.2, if *a* increases with time, that is, if the universe expands, the distance between objects in the universe increases.

Galaxies will recede from us with velocity v given by the formula

$$v = HR$$

where R is the distance of the galaxy from us and H is defined by

$$H = \frac{1}{a}\frac{da}{dt}$$

The parameter H is known as *Hubble's constant* and the formula for the recession velocity of galaxies is known as *Hubble's law,* named after Edwin Hubble, the astronomer who first observed this behavior. Actually, the term *constant* is misleading, since H changes with time, although its present change on a human time scale is negligible.

The velocity of recession, of course, cannot be measured directly; it is deduced from the shift toward lower frequencies (red shift) of the spectral lines of the light emitted by the object. In order to observationally determine whether Hubble's law is satisfied, we also need a means of obtaining the distance R to faraway objects. The method for doing this is based on the following simple fact. If one knows the intrinsic brightness of an object, one can determine how far away it is by how bright it appears to be. The problem is to find distant "standard candles," the intinsic brightness of which are known. The brightest galaxies in a cluster of galaxies have been used as standard candles, because we believe that their intrinsic luminosity does not vary greatly from cluster to cluster. The determination of the value of their intrinsic brightness involves a rather complicated scheme in which the intrinsic luminosity of nearby objects is used to calibrate the intrinsic luminosity of objects which lie father away. Thus, one starts with nearby stars in our own galaxy, the distance (and hence luminosity) of which can be determined by direct means. From the correlation obtained between intrinsic luminosity and spectral types of the stars, one can then determine the distance of star clusters in our galaxy. This information is then used to determine the intrinsic brightness of Cepheid variable stars. These stars are bright enough to be seen in nearby galaxies, and thus the distance of these galaxies is determined. This information is used to calibrate the intrinsic luminosity of the brightest objects in these galaxies. These brightest objects are then used as a standard candle to determine the distance of nearby clusters of galaxies. Finally, from the distance of the nearby clusters of galaxies, we can calibrate the intrinsic luminosity of the brightest galaxies in a cluster.

When one makes a graph of the distance to a cluster of galaxies (determined in the above manner) versus its red shift one finds that the data points are very well fit by a straight line, in accordance with Hubble's law. The slope of this line determines Hubble's constant H. In this manner, H is found to be about 30 miles per second per megaparsec (where one megaparsec is about three million light years, nearly twenty billion billion miles). In other words, an object lying one megaparsec away from us will have a recession velocity due to cosmological expansion of 30 miles per second. Because of possible errors and uncertainties in the distance calibration procedure described above, this value of H quite possibly could be wrong by as much as a factor of two. However, the validity of Hubble's law is very well established.

The *deceleration parameter q* is defined by

$$q = -a\frac{d^2a}{dt^2} \Big/ \left(\frac{da}{dt}\right)^2;$$

q tells us how much the expansion of the universe is slowing down (or speeding up). In principle, q can be determined observationally as follows: The light travel time from very distant objects is so great that the red shift of their spectral lines is significantly affected by the expansion of the universe at earlier epochs, not just by the present expansion rate. Hence, if q is nonzero, that is, if the expansion rate is changing, one will observe deviations from Hubble's law for very distant objects. From these deviations from Hubble's law, the value of q can be deduced. As discussed further in 5.2, such observations are very difficult to make.

In terms of H and q, the equations of 4.2 become

$$q = \frac{4\pi G}{3H^2} (\rho + 3P/c^2)$$

$$H^2 = \frac{8\pi G}{3} \rho - \frac{kc^2}{a^2}$$

4.5   How Seriously Should We Take the Prediction of a Spacetime Singularity?

The theoretical prediction described in 4.3 that the universe must have begun a finite time ago with a "big bang" spacetime singularity and that if it is closed it will end with a similar singularity are certainly very astounding and exotic ideas. The observational evidence leaves little, if any, doubt that the universe is dynamic and has expanded from a much more condensed state, but how seriously should we take the prediction that the matter density of the universe actually was infinite in the past? There are two senses in which we may ask this question: (1) Is it an inescapable conclusion that general relativity predicts a spacetime singularity? (2) If so, can we believe the predictions of general relativity in such extreme circumstances or do we expect the theory to break down near spacetime singularities?

By question 1, we mean the following: The equations and conclusions obtained in 4.2 and 4.3 were all based on the assumption that the universe is *exactly* homogeneous and isotropic. However, we know that this assumption is only approximately valid; it is only when we average over very large distance scales that the universe even begins to appear homogeneous and isotropic. Is it possible that the prediction of a spacetime singularity is purely a consequence of this mathematical assumption of exact symmetry; that if we were only to assume approximate symmetry we would no longer predict singularities? In particular, consider the case of the closed universe. As mentioned in 4.3, under the assumption of exact homogeneity and isotropy general relativity predicts that the universe will recollapse to a singular state. Suppose, now, that the assumption of *exact* symmetry is dropped. Is it possible that—rather than recollapse to a singular state—the universe will recollapse to a very condensed state but then "bounce" and reexpand again? Rather than having begun a finite time ago in the past and ending a finite time from now in the future, perhaps the history of the universe consists of an infinite cycle of phases of expansion and recontraction.

Before the mid 1960s, many theorists believed that the prediction of singularities is merely a consequence of the assumption of exact symmetry and that the "bouncing" universe model is not only consistent with general relativity but is the correct model of our universe. However, within the past ten years, a number of results have been proven (mainly by Roger Penrose and Stephen Hawking) which show that these beliefs are wrong. In other words, the answer to question 1 above is yes; the "bouncing" universe model is not in accord with general relativity.

As mentioned in 3.4, it is very difficult to find exact solutions of Einstein's equation. Thus, the investigation of the occurrence of singularities in general relativity has followed the path of proving general theorems concerning the behavior of solutions. The main result in this line of investigation can be stated as follows.

Singularity Theorem. Hypothesis: Assume general relativity is correct and suppose further that the following four conditions are satisfied: (1) The mass density and pressure of matter never become negative. (2) There are no closed timelike curves in the universe, that is, it is impossible to visit one's own past (within the framework of general relativity, such an exotic possibility can occur). (3) Either (*a*) the universe is closed, or (*b*) a condition is satisfied which basically says that there is enough matter in the universe to refocus light via "light bending." (4) A relatively technical (but very reasonable) mathematical condition is satisfied.

Conclusion: There must be at least one timelike or null geodesic in the universe which has only finite "length" in the future or in the past.

Recall that timelike geodesics represent the paths in spacetime of freely falling observers and null geodesics represent the paths of light rays (see 2.7 and 3.3). The theorem states that if the above four very reasonable conditions hold, then—even if all assumptions about homogeneity, isotropy, or other symmetries are dropped—the history of at least one freely falling observer or light ray must come to an end within a finite time in the future or must have begun a finite time ago in the past (or both). In the homogeneous and isotropic case, the history of *all* freely falling observers began a finite time ago in the big bang singularity. Thus, the spacetime singularity required by the theorem is not as all-encompassing as the big bang singularity. Furthermore, the theorem does not say that the density of matter or the spacetime curvature must be infinite where the history of the observer began or ended. However, the theorem does establish the fact that the occurrence of spacetime singularities is a true feature of general relativity, not merely a consequence of convenient mathematical assumptions concerning symmetries. The theory of the nature of the possible types of spacetime singularities in general relativity is currently an active topic of research.

Let us turn to question 2 stated at the beginning of this section: Do we expect the theory of general relativity to break down in the extreme conditions near a spacetime singularity? The answer is yes. We know that on a microscopic scale, nature is governed by the laws of quantum theory. However, the principles of quantum mechanics are not incorporated into general relativity. Hence, we do not believe that general relativity can be a true, final theory of nature. Classical mechanics (that is, Newton's laws of motion) provides us with an accurate description of the motion of macroscopic bodies, but it breaks down when we attempt to apply it on atomic distance scales. In a similar manner, we believe that general relativity provides an accurate description of our universe under all but the most extreme circumstances. However, near the big bang singularity when the scale factor a goes to zero and the density and curvature become infinite, we expect general relativity to break down.

What is the new, fundamental theory of nature which incorporates the principles of both general relativity and quantum theory? What does this theory say about spacetime singularities? Even the most optimistic theorist can only hope for the beginning of an answer to these questions within the foreseeable future.

Five

The Evolution of Our Universe

5.1 The Evolution of the Universe to Its Present State

In this section we shall outline our best current picture of the history of our universe from the big bang to the present. The assumptions which go into this picture are (1) Einstein's theory of general relativity with no cosmological constant term, (2) homogeneity and isotropy of the universe, (3) all known physical properties of matter, and (4) current observational evidence. Some of the quantitative details in the picture given below may change as better observational data become available. (All the numbers should be taken as approximate in any case.) However, the basic qualitative picture is not very sensitive to the fine details of the observational data. The discussion applies to both an open or a closed universe.

We label each epoch in the history of the universe by the time t as measured by the clock of a freely falling (geodesic) observer to whom the universe appears homogeneous and isotropic. The world lines of our galaxy and of other galaxies are presently good approximations to the world lines of such observers.[1] Here, then, is the history of our universe.[2]

$t = 0$; The Big Bang

1. On the other hand, the universe would not appear isotropic to observers in our vicinity who move at velocity near the speed of light with respect to us, since, for example, their relative velocity with respect to other galaxies would depend strongly on direction. It is not difficult to show that by the twin paradox type effect (2.6) such observers would always measure a smaller elapsed time between the various epochs in the history of the universe.

2. The symbol $\lesssim$ used below should be read as "less than or approximately equal to"; the symbol $\gtrsim$ means "greater than or approximately equal to." The number 10^{-43} means a decimal point followed by 42 zeros followed by a 1 (a very small number); the number 10^{92} means a 1 followed by 92 zeros (a very large number).

According to the prediction of general relativity, the matter density and spacetime curvature of the universe were infinite.

$t \lesssim 10^{-43}$ Seconds

Dimensional arguments indicate that this close in time to the big bang singularity, general relativity should not be valid and should be replaced by a new theory which incorporates the principles of quantum theory (see the end of 4.5). In other words, we don't know anything about this epoch.

10^{-43} Seconds $\lesssim t \lesssim 1$ Second

During this period the universe expanded and the matter density dropped from about 10^{92} times the density of water (!) to about 500,000 times the density of water. During the early part of this period, the matter in the universe consisted primarily of very high-energy elementary particles in thermal equilibrium. Although we can be reasonably confident that general relativity provides a correct description of the universe in this and all later epochs, our knowledge of high-energy elementary particle physics is not good enough to allow us to give a detailed description of what occurred in this period. However, by the end of this period, the temperature of the matter cooled to about 10 billion degrees centigrade (!) and all the exotic elementary particles decayed. The matter in the universe by the end of this period consisted of a "soup" predominantly composed of photons (quanta of electromagnetic radiation), neutrinos (elementary particles which interact very weakly with matter and which —like the photon—have no rest mass and travel at the speed of light), electrons and their antiparticles (positrons), protons, and neutrons. The presence of protons is favored over the presence of neutrons because protons are slightly less massive; at the end of the period there were about five times as many protons as neutrons.

1 Second $\lesssim t \lesssim 1,000$ Seconds

The universe continued to expand; the density of the soup dropped from about 500,000 times the density of water to about half the density of water; the temperature dropped from about 10 billion degrees to about 1 billion degrees. Early in this epoch, the positrons combined with electrons, converting their mass energy to photons.

But the most dramatic thing that took place during this period was nucleosynthesis: the protons and neutrons underwent nuclear reactions and formed elements. Prior to this period the temperature of the matter was too high and any elements which might have been formed would have been dissociated immediately. After this period, the temperature and density of the matter was too low to permit these nuclear reactions to take place. Only during this period lasting about 15 minutes were conditions right for nucleosynthesis. Most of the neutrons present at the beginning of this period combined with the protons to form helium nuclei. (The neutrons which did not react decayed into protons and electrons either during this period or very shortly thereafter.) A very small amount of deuterium (that is, "heavy hydrogen," a proton and a neutron bound together) was also synthesized, but essentially no other element formation occurred. Thus, at the end of this process, about 25% by mass of the original protons and neutrons was converted into helium, while essentially all the rest was left in the form of hydrogen (that is, protons). *This percentage of helium is observed in cosmic abundances.* Thus, nucleosynthesis in the early universe accounts for most of the helium we presently find in the universe. (Nucleosynthesis in stars accounts for the remainder of the helium and for the presence of heavier elements; see chap. 6.) The percentage of helium synthesized is sensitive to the original ratio of neutrons to protons but it is not very sensitive to the magnitude of the density of neutrons and protons, since nearly all the neutrons react in any case. However, the percentage of deuterium synthesized depends strongly on the neutron-proton density. The higher this density the more likely it is that the deuterium produced will react further to form helium, and the less total deuterium will result at the end of the process.

1,000 Seconds $\lesssim t \lesssim$ 100,000 Years

The universe continued to expand rapidly and cool, but otherwise this was a relatively dull period. The important constituents of the "soup" of matter and radiation filling the universe during this period were photons, protons, helium nuclei, and free electrons. The neutrinos mentioned above were still present during this period and, indeed, these primordial neutrinos can presumably be found in the present universe, although by now their energy has been greatly reduced by the expansion of the universe and detection would be extremely difficult. However, the neutrinos interact so weakly with other forms of matter that they have no important direct effects on other matter during this period or later. Nevertheless, during this period the neutrinos made a nonnegligible contribution to the energy density of the universe. Free electrons and hydrogen and helium ions—rather than neutral hydrogen and helium atoms—were present during this period because the temperature was still high enough to keep those atoms ionized. Photons interact with charged particles such as electrons and hydrogen and helium ions; charged particles can absorb, emit, or scatter photons. In this sense, during this period the electromagnetic radiation (photons) was strongly coupled to the matter (electrons and hydrogen and helium ions). The radiation and matter were in thermal equilibrium with each other. By the end of this epoch, the temperature of this soup of radiation and matter had dropped to about 4,000 degrees centigrade.

$t \sim 100,000$ Years

As the temperature dropped to 4,000 degrees and below, another very important effect took place: the electrons combined with the hydrogen and helium ions to form neutral hydrogen and helium atoms. When this process was completed, there were no longer any free charged particles. As a result, the interaction between photons and matter was very greatly reduced. The electromagnetic radiation decoupled from the matter and they no longer remained in thermal equilibrium with each other.

$100,000$ Years $\lesssim t \lesssim 8$ to 18 Billion Years
(Present Era)

The photons cooled without significant interference from the matter as the universe continued to expand. By the present era this primordial electromagnetic radiation—which fills the entire universe—has cooled to about 3 degrees Kelvin (that is, 3 degrees above absolute zero), which puts it in the microwave range. *This cosmic microwave radiation has been observed.* The discovery of it was made in 1965 rather by accident: A.A. Penzias and R.W. Wilson, working at Bell Laboratories, found what they at first thought was excess background noise in a large radio antenna built to observe the Echo satellite. They and a group at Princeton University headed by Robert Dicke soon identified this "noise" as the radiation originating from the early universe. More recent observations have demonstrated the thermal nature of this radiation. Furthermore, the radiation has been found to be isotropic to an extremely high degree of accuracy. The existence of this radiation is a striking confirmation of the picture presented here of the history of the universe; it is the strongest evidence we have that the universe has expanded from a much denser and hotter phase. Furthermore, the observed isotropy of the radiation is the best piece of evidence we have that the universe is very nearly homogeneous and isotropic on large distance scales, at least from time $t \sim 100,000$ years onward.

In the meantime, at the beginning of this period, small inhomogeneities in the matter distribution began to grow under their own self-gravitational attraction. Prior to this period, the growth of most inhomogeneities was inhibited by the radiation pressure on the matter. However, the decoupling of matter and radiation at $t \sim 100,000$ years permitted the growth of inhomogeneities of galactic and smaller mass. Thus, the galaxies, star clusters, and stars condensed out of the primordial medium. However, our understanding of the details of the process of galaxy formation is not very good; this is presently an active area of research. In particular, we do not have a good understanding of why the masses of galaxies are typically one hundred billion (10^{11}) times the mass of our sun, rather than much larger or much smaller.

Finally, we should mention that some 3.5 to 13.5 billion years after the big bang our sun and solar system formed. Another 4.5 billion years more brings us to the present time.

5.2   The Future Evolution of Our Universe

We now have described the evolution of the universe from the big bang to its present state. What is going to happen in the future? (Of course, we are talking in terms of time scales of tens of billions of years.) The most important question in this regard is, Is the universe open or closed? As discussed in 4.3, if the universe is open ($k = 0, -1$) it will continue to expand forever. If the universe is closed ($k = +1$), the expansion will eventually come to a halt; the universe will recontract and will again approach a singular state within a finite time.

In the present universe, the pressure, P, of matter is negligible. Hence, the equations of 4.4 become

$$q = \frac{4\pi G\rho}{3H^2}$$

$$H^2 = \frac{8\pi G\rho}{3} - \frac{kc^2}{a^2}$$

Below, we shall describe four observational methods for determining whether our universe is open or closed.

1. *Measure the mass density of the universe:* According to the second equation above, if the present density of the universe is greater than $3H^2/8\pi G \sim 5 \times 10^{-30}$ gm/cu cm, the universe must be closed (k positive). If $\rho \leqq 3H^2/8\pi G$, the universe must be open (k zero or negative). Thus, a determination of the density of matter in the universe will tell us if the universe is open or closed.

The best estimate of the mass density of the universe due to matter in galaxies is about 2.5×10^{-31} gm/cu cm. This is some twenty times less than the mass density required for a closed universe. The galactic mass density is, of course, only a lower limit to the true mass density of the universe. We know that there must be at least that much mass density but there might be more that we do not see. However, the most plausible forms in which significant amounts of matter might be hidden—for example, in a tenuous hydrogen gas between the galaxies—seem to be ruled out by a combination of observational and theoretical arguments. Thus, the observation of the mass density of the universe provides fairly strong evidence that the universe is open.

Incidentally, the measurements of Hubble's constant H (see 4.4) and the mass density ρ both require a determination of the distance to faraway objects. As discussed in 4.4, distances are determined by a rather complicated scheme, and there is plenty of opportunity for errors to be made. However, it turns out that the ratio of ρ to H^2— that is, the quantity relevant to the determination of whether the universe is open or closed—is relatively unaffected by any such errors. Thus, an error in the distance scale of a factor of two (which is certainly in the realm of possibility) would result in an error of a factor of two in the determination of H and an error of a factor of four in the determination of ρ, but would leave the ratio ρ/H^2 unchanged.

2. *Measure q:* As mentioned in 4.4, in principle, the deceleration parameter q can be measured by observing the deviation from Hubble's law of very distant objects. According to the two equations given above, if $q > 1/2$ then k must be positive and hence the universe must be closed; on the other hand, if $q \leqq 1/2$, then the universe must be open. However, in practice many difficulties arise when one attempts to determine q. In the first place one must look at objects which are so far away that they can barely be seen with the best telescopes. Thus, the observations are extremely difficult to make. But there are other difficulties which are even more serious. As discussed in 4.4, the determination of distances (required to establish Hubble's law or measure the deviation from it) is based on the hypothesis that the intrinsic luminosity of the brightest galaxies in a cluster of galaxies do not vary significantly from cluster to cluster. This appears to be true for reasonably nearby clusters of galaxies. However, when we look at very distant galaxies, because of the long light travel time we see them as they were at a much earlier phase of their evolution. Unfortunately it seems very likely that the intrinsic luminosity of a galaxy varies significantly during its evolution. Thus, we cannot be certain of the intrinsic luminosity of the most distant galaxies we observe. This means that at present we cannot reliably determine by this method the distances of the far-away objects which are required to measure q. At present, the only firm conclusion one can draw is that q cannot be very large (for example, much greater than 2) since if it were one would observe deviations from Hubble's law for relatively nearby objects.

3. *Determine the age of the universe:* In a closed universe, the rate of expansion decreases more rapidly than in an open universe (that is, as mentioned above, q is larger for a closed universe). This means that if the universe were closed, less time would elapse before the expansion would slow down to its presently observed value than if the universe were open. In other words, by measuring the present age, t, of the universe we can, in principle, determine if the universe is open or closed. In fact, it is not difficult to show that if $t < 2/3H$

then the universe must be closed, while if $t \geqq 2/3H$ the universe must be open.

The present age of the universe can be estimated from the age of elements and the age of the oldest observed star clusters. These observations indicate that the universe must be at least about 8 billion years old but not older than about 18 billion years. Using the value of Hubble's constant given in 4.4, the critical age $2/3H$ is about 13 billion years; this number, however, should not be taken too literally, since the actual value of H is rather uncertain because of possible errors in determining the distance scale. Thus, the observed age of the universe is in rough agreement with the critical age $2/3H$ determined from the dynamics of the universe—a fact which reinforces one's confidence in the validity of the basic picture of the evolution of the universe given above. However, at present the uncertainties in both the observed age and the value of the critical age are too great to allow us to draw any conclusions by this approach as to whether the universe is open or closed.

4. *Measure the cosmic deuterium abundance:* As mentioned above, if the universe is closed, the present matter density of the universe must be relatively large (greater than the critical density $3H^2/8\pi G$). If this is the case, then the density of protons and neutrons during the phase of nucleosynthesis in the early universe also must have been relatively large. In that case, as already mentioned in 5.1, the percentage of the original protons and neutrons which end up as deuterium will be relatively small, since almost all the deuterium which is synthesized will be converted to helium. Considerably more deuterium will be produced if the universe is open. Thus, by measuring the abundance of deuterium in the universe we learn a great deal about whether the universe is open or closed. Within the past several years, observations have been made of the fraction of cosmic matter in the form of deuterium. The abundance of deuterium has been found to be high, thus providing strong evidence that the universe is open.

In summary, based on the presently available observational evidence, the best guess is that the universe is open ($k = -1$; hyperboloid geometry) and will, therefore, continue to expand forever. However, there is still considerable room for doubt in this conclusion. Of the four pieces of evidence discussed above, two of them (2 and 3) are inconclusive while many possible (though not necessarily plausible) loopholes remain in the arguments based on the other two. For example, the existence of a number of possible forms of unseen matter has not been ruled out, thus providing a loophole for the arguments based on observations of density; in the deuterium evidence the high abundance of deuterium could conceivably have resulted from its being manufactured by some process other than nucleosynthesis in the early universe. Perhaps within the next few decades, further observations will be made which will conclusively tell us the eventual fate of our universe.

Six

Stellar Evolution

In this chapter we shall focus our attention on the formation, evolution, and eventual fate of stars. From the grand viewpoint taken in the last two chapters, the evolution of stars might appear to be a relatively minor phenomenon in the evolution of the universe, but from any other viewpoint it is of considerable importance and interest. However, the real reason for discussing stellar evolution here is to set the stage for the discussion of gravitational collapse and black holes, which will be the subject of the remainder of this book.

Except for the final stages of stellar evolution (6.5), the gravitational field of a star is sufficiently weak that the Newtonian theory of gravity applies. Thus, in this chapter we shall describe phenomena in ordinary Newtonian language rather than the much more abstruse language of general relativity.

6.1   Stellar Evolution: Birth

The process of the formation of a star begins when a cloud of gas starts to contract under its own self-gravitational attraction. As it contracts, gravitational potential energy is converted to thermal energy and the gas cloud heats up. As it heats up, the pressure of the gas builds up, tending to halt the collapse. If the gas cloud did not lose energy, this pressure would entirely stop the collapse at an early stage. But the gas cloud does lose energy. Since its temperature is high, light and other forms of electromagnetic radiation are radiated away from its outer surface. As a result, the gas cloud cannot maintain the required pressure, and the cloud continues to collapse slowly. As it collapses, it continues to get hotter and hotter.

After several million years of this slow collapse, the center of the gas cloud becomes sufficiently hot and dense that nuclear reactions begin to occur. Hydrogen is converted to helium, releasing considerable energy. (This is the same basic nuclear fusion process as occurred in the early universe [5.1] and as hopefully will be employed here on earth in the future as a source of energy; it is also the same basic process as occurs in the hydrogen bomb.) At this point the gas cloud is stabilized. The energy radiated away from the surface is now balanced by the nuclear energy generation, so the cloud does not have to collapse further to gain the thermal energy required to maintain the pressure needed for self-support. We now have a star.

6.2   Continued Evolution

Once formed, the star enters a long, stable phase lasting billions of years during which hydrogen is converted to helium at the center of the star. Interestingly, the more massive the star, the shorter this phase, since in massive stars the hydrogen is "burned" to helium at a much faster rate. Our sun was formed about 4.5 billion years ago and will continue in this stable phase of evolution for probably another five billion years.

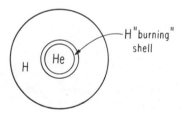

Figure 19   The configuration of a star (not drawn to scale) after billions of years of nuclear reactions at the center converting hydrogen (H) to helium (He).

Eventually, a large core of helium is built up at the center of the star. Hydrogen is still being "burned" in a layer outside the helium core, and the structure of the star is as pictured in figure 19. Now the helium core of the star finds itself in much the same situation as the original gas cloud. It begins to contract due to the force of gravity and it heats up. This causes the hydrogen burning in the layer outside the core to proceed at a faster rate. As this happens, the outer layers of the star expand and cool. This causes the appearance of the star to change significantly. The star becomes a *red giant*—"red" because the surface temperature is low; "giant" because of the great expansion of the outer layers. Some mass loss may take place during this phase, that is, some of the outer layers of the star may get entirely blown off. Our sun will become a red giant in about 5 billion years.

Soon the center of the helium core becomes so hot and dense that nuclear reactions begin to occur there, converting the helium primarily to carbon and oxygen. Energy is released in this process and the collapse of the core is again (temporarily) stabilized.

6.3   Permanent Stabilization against Collapse? White Dwarfs

We have seen above that stars can be supported against collapse if they are provided with an energy source such as nuclear burning. However, the energy resources of a star are finite and eventually they will be depleted, so support by a means requiring an energy source can only be temporary. Can a star ever be permanently supported against gravitational collapse?

Stars are composed of hydrogen, helium, and other elements manufactured by nuclear reactions. The temperature of a star is high enough so that the atoms are ionized. If the density is not extremely high, this collection of ions and electrons behaves like an ideal gas, that is, its pressure, P, is related to its temperature, T, and number density, n, by the formula

$$P = nkT$$

where k is Boltzmann's constant. For a star composed of an ideal gas the answer to the above question about self-support is no; permanent support against collapse is impossible. From the above formula one needs large thermal energy (large T) to obtain large values of pressure at a given density. Since stars radiate energy from their surface, they cannot maintain the pressure required for self-support except by an additional energy source.

However, stars are *not* composed of a perfect, ideal gas. Electrons obey the Pauli exclusion principle of quantum mechanics: no two electrons can occupy exactly the same state. As a consequence of this fact, there is an additional pressure exerted by the electrons. Under ordinary conditions, this quantum mechanical effect is negligible, but at very high densities it is very important. When the density of matter in a star becomes greater than about five million times the density of water, the electrons contribute a pressure P due to this effect given by

$$P \sim hcn^{4/3}$$

where h is Planck's constant, c is the speed of light, and n is the number of electrons per unit volume. (At somewhat lower densities P is proportional to $n^{5/3}$.) This is known as *electron degeneracy pressure*. The important thing about electron degeneracy pressure is that it does not require the presence of large thermal energy (high temperature), so it can be maintained even as the star radiates its energy.

How much pressure is needed to support a star against collapse? The requirement for hydrostatic equilibrium of a star is that at each radius r the Newtonian gravitational force be balanced by the force arising from the pressure of the gas. The equation which expresses this requirement is

$$\frac{dP}{dr} = -G\,\frac{m(r)}{r^2}\,\rho.$$

Here the left side of the equation represents the pressure gradient, that is, the rate of change of pressure with respect to radius; ρ is the mass density and $m(r)$ is the total amount of mass enclosed by a shell of radius r. From this equation it is not difficult to show that in order to support itself against collapse, the pressure P_c at the center of the star must be roughly

$$P_c \sim GM^{2/3}\rho^{4/3}$$

where M is the total mass of the star.

To see if electron degeneracy pressure can support a star, we must compare the pressure it yields to the central pressure needed for support. As the above approximate formulas show, the pressure obtainable at high densities and the pressure needed for support have the same dependence on electron number density n (since the mass density ρ is proportional to n). Support will be possible when the coefficient of $n^{4/3}$ in the formula for electron degeneracy pressure is greater than the coefficient of $n^{4/3}$ in the formula for the central pressure needed. Since this latter coefficient depends on the mass M of the star, this means that stars with sufficiently small mass can be supported, while stars with large mass cannot. If P were exactly proportional to $n^{4/3}$, even the stars with sufficiently low mass would be unstable. However, since P is actually proportional to $n^{5/3}$ at lower densities, stable configurations do exist below the critical mass.

Detailed calculations, first performed by S. Chandrasekhar, show that for stars with mass less than about 1.3 times the mass of the sun, electron degeneracy pressure permanently halts collapse. For stars with mass less than about half a solar mass this happens before the phase of helium burning mentioned above can occur; for stars with greater mass (but less than 1.3 solar masses) it occurs at a later stage of evolution. Stars that are supported by electron degeneracy pressure are called *white dwarfs*—"dwarfs" because they are very condensed and thus small in size; "white" because their surface temperature is very high when they enter this phase. No further evolution occurs for white dwarfs; they simply cool down forever. Many stars in our vicinity are white dwarfs. Our sun will end its life as a white dwarf.

On the other hand, if M is greater than about 1.3 times the mass of the sun, electron degeneracy pressure is not sufficient to support the star. Some stars whose initial mass is greater than 1.3 solar masses may be able to shed enough mass during evolution (for example, during the red giant phase) so that they nevertheless end up as white dwarfs with $M < 1.3$ solar masses. However, if this does not take place, further evolution of the star must occur.

6.4   Final Stages of Stellar Evolution; Supernova

Let us now continue our description of the history of a star whose mass is greater than 1.3 solar masses. At the end of 6.2, the large helium core at the center of the star had become so dense and hot that nuclear reactions resulting in the production of carbon and oxygen began to take place at the center of the core. The basic scenario which previously occurred now repeats itself: A core of carbon and oxygen is built up at the center of the helium core. This core contracts and heats up until the carbon begins to undergo nuclear fusion reactions, producing neon and other elements.[1]

The basic pattern of core buildup, followed by contraction and heating, followed by nucleosynthesis of new elements, continues to repeat itself. In each case, the new nuclear reactions provide the thermal energy required in order to get enough central pressure to support the star. Eventually, the configuration of the star will be roughly as follows: at the center of the star will be a gaseous core of iron and nickel, surrounded by successive layers of lighter elements; the outer envelope of the star will still be composed predominantly of hydrogen. The central iron-nickel core can be supported by electron degeneracy pressure until its mass becomes too great. At this point, the star reaches the end of the line in its ability to support itself. Iron and nickel are the most stable of all nuclei, so no further energy generating nuclear reactions can occur in the core.

Thus, the central core is unable to support itself by any means and undergoes gravitational collapse. As it contracts, the core heats up to enormous temperatures and photodisintegration takes place. The high-energy photons which are present break apart the nuclei back into their elementary constituents, that is, neutrons and protons. When the density of the collapsing core becomes about 200 billion times the density of water (!) another very important process takes place at a rapid rate: electrons combine with protons to produce neutrons and neutrinos. (At such enormously high densities, conditions are favorable for an electron and proton to "fuse" together to form a neutron.)

As mentioned in chapter 5, the neutrino is an elementary particle which—like the photon—has no rest mass and travels at the speed of light. Under ordinary circumstances, the interaction of neutrinos with matter is extremely weak and it is very difficult to even detect the presence of a neutrino. However, under the conditions considered here—where the neutrino energy and matter density are both very high—it is believed that the neutrinos have an extremely

---

1. Some calculations indicate that in some cases this carbon burning may occur so explosively that the entire star is blown apart! This phenomenon— if it occurs at all—is known as a *carbon detonation supernova*. Carbon detonation supernovas cannot occur if the mass of the star is sufficiently large, that is, greater than about seven solar masses.

important effect. A current hypothesis is that at least in some cases the neutrinos emitted from the reactions in the collapsing core may "blow off" the outer layers of the star. This is believed to provide the explanation of *supernovas,* the disruptive explosions of stars, several of which have been seen in past centuries in our own galaxy and many more of which have been observed in other galaxies. At present, no one has quite succeeded in getting this explanation of supernovas to work in detail in a calculation, but it is widely believed that the basic idea is correct. This is presently an active area of research.

We should mention one important by-product of stellar nucleo-synthesis and supernovas. They explain the observed cosmic abundances of the elements. As described in chapter 5, after the nuclear reactions which take place in the early universe one has only hydrogen and helium; essentially no heavier elements are produced. Where, then, did the carbon, oxygen, iron, and all the other elements we observe come from? The answer is that these elements were formed by nuclear reactions in stars and were then disbursed throughout the galaxy by supernova explosions. (Some heavy elements, like uranium, are probably manufactured by nuclear reactions occurring during the extreme conditions of the supernova explosion itself.) Our sun and solar system condensed out of matter which already had been enriched by the nucleosynthetic processes of an earlier generation of stars.

6.5 The Collapsing Core; Neutron Stars

At this final stage of stellar evolution, the core of the star is composed primarily of neutrons and is undergoing rapid gravitational collapse. What happens next? In the first place, the gravitational field of the core is no longer weak; the Newtonian theory of gravitation is no longer a valid approximation and one must use the full theory of general relativity. The Newtonian equation for hydrostatic equilibrium given in 6.2 must be replaced by the exact equation

$$\frac{dp}{dr} = G \frac{(\rho + p/c^2)\ (m(r) + 4\pi r^3 p/c^2)}{r(r - 2Gm(r)/c^2)}.$$

Note that the right side of this equation is always larger than the right side of the corresponding equation in 6.2. Thus, in general relativity it is *harder* to maintain hydrostatic equilibrium, that is, more pressure is needed to support a star, than in Newtonian theory.

Calculations show that if the mass of the collapsing central region of the star is less than about 2 solar masses, it should be possible for *neutron degeneracy pressure* to support it. (Neutron degeneracy pressure is the analogue of electron degeneracy pressure discussed in 6.3.) If this is the case the collapse can be halted[2] and one can end up with a stable configuration which—like a white dwarf—simply cools down forever. Unlike a white dwarf, this end product of stellar evolution will be composed of neutrons at truly enormous densities, more than one hundred trillion (10^{14}) times the density of water. This exotic object is called a *neutron star*. Since the density of matter in a neutron star is comparable to the density of matter in an atomic nucleus, one can think of a neutron star as an enormous nucleus bound together by gravitational rather than nuclear forces. A typical radius of a neutron star is only about 7 miles. Because of uncertainties in our knowledge of the behavior of matter at such enormous densities the precise value of the maximum possible mass of a neutron star (about 2 solar masses) is not as well established as the value of the maximum possible mass of a white dwarf (1.3 solar masses).

Remarkably, the existence of neutron stars has been confirmed recently by the discovery of pulsars in 1967. Pulsars are astronomical objects which emit electromagnetic radiation in regular clockwise pulses with periods ranging from several hundredths of a second to several seconds. By far the most plausible explanation of pulsars is that the electromagnetic radiation comes from some sort of "hot spot" on a rotating body. During each revolution of the body the hot spot gets pointed toward us and we see a signal. This signal is exactly reproduced in the succeeding revolutions. However, we know of no astronomical body that can rotate (or even vibrate)

---

2. When the collapse of the core is halted, a shock wave will be produced. It is possible that this shock wave may cause the outer layers of the star to be expelled. If so, this would provide an explanation of supernovas alternative to the above-mentioned hypothesis that neutrinos are responsible for "blowing off" the outer layers. Our present knowledge of the detailed dynamics of collapse and supernovas is very rudimentary, and a great deal more research in this area needs to be done.

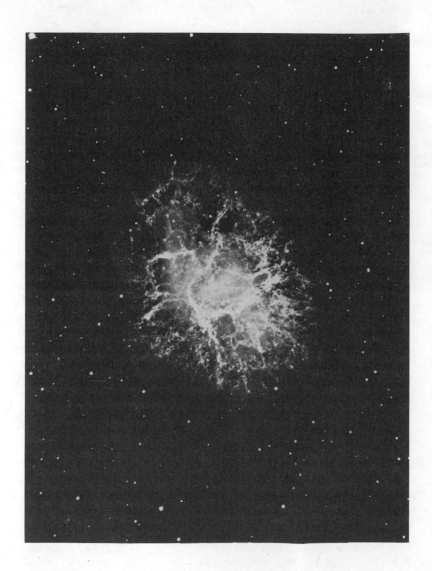

Figure 20   The Crab nebula. (Yerkes Observatory photograph.)

with such a small period except a neutron star; any body which is less compact would fly apart if it were rotating at that rate. A satisfactory detailed model of the hot spot mechanism by which neutron stars emit radiation has not yet been given, but the idea that pulsars are rotating neutron stars is now generally accepted.

If the ideas described in this section and the preceding one are correct, we should expect that in some cases a supernova explosion will leave behind a neutron star remnant, and that in some cases this neutron star will act as a pulsar. Are any pulsars found at the site of previous supernova explosions? Yes. The most dramatic example is the Crab nebula (see fig. 20). According to ancient Chinese records an enormous brightening of a star in that part of the sky took place in July of A.D. 1054. Today we see there a gas cloud expanding outward at high velocity. Clearly this gas cloud was the outer layer of a star that underwent a supernova explosion in 1054. Finally, at the center of this gas cloud we find a pulsar.

In summary, stars with mass less than about 1.3 solar masses will end their lives peacefully as white dwarfs. If the mass of a star is too great to allow this option, it will undergo further evolution which will eventually lead to the collapse of the central core of the star and (at least in some cases) a supernova explosion in which the outer layers of the star are blown off. If the mass of the collapsing core is less than about 2 solar masses, its collapse may be halted when nuclear densities are reached, thus producing a neutron star. However, what if the mass of the collapsing core is greater than about 2 solar masses, *or* the infall velocity is too great for the collapse to be halted even if the mass is somewhat smaller than 2 solar masses, *or* what if additional mass later falls on a neutron star or white dwarf raising its mass above the critical value for which support is possible? In such cases all possible opportunities—temporary and permanent—for stabilizing the star against collapse have been used up. Complete gravitational collapse must occur!

Seven

Gravitational Collapse to Black Holes

At the end of chapter 6 we concluded that in the final stage of evolution of a sufficiently massive star, the central core of the star rapidly undergoes collapse and has no possibility of ever halting this collapse. What happens when a body undergoes this kind of catastrophic gravitational collapse? What becomes of the body and its gravitational field following collapse? In this chapter we will outline the chain of reasoning which has led theorists to conclude that the complete gravitational collapse of a body always results in a black hole and that the detailed nature of the black hole thus produced depends only on its mass, angular momentum, and net electric charge.

7.1   Spherical Collapse; Black Holes

Unfortunately, the theoretical analysis in general relativity of what happens during gravitational collapse in general situations—even ignoring the complexities of the behavior of matter—is extremely difficult to perform. It is extremely difficult to find exact solutions of Einstein's equation which describe gravitational collapse; even attempts to find approximate solutions by numerical integration using a computer have not yet been successful, though further research in this area is now being actively pursued. Fortunately, this situation changes significantly for the better when one makes the simplifying assumption of exact *spherical symmetry,* that is, if one assumes that the body and its gravitational field are spherical in shape, or, in other words, that everything "looks the same" when rotated by an arbitrary angle about a central point (the center of the body). In the spherically symmetric case, the equations become tractable and one can analyze in detail what happens during collapse.

In ordinary flat spacetime (Euclidean geometry) the surface area

A of a sphere is related to its radius R (distance from the origin) by the formula

$$A = 4\pi R^2.$$

However, in general relativity the spacetime geometry in the vicinity of a collapsing body will be significantly curved and this relation will no longer hold. Nevertheless, it is very convenient to label the spheres in a curved spherically symmetric spacetime by the parameter R defined by the above formula (that is, $R = \sqrt{A/4\pi}$). R is still called the radius of the sphere, even though it is no longer necessarily equal to the distance of the surface of the sphere from the origin. With this convention understood, we shall now describe the salient features of what happens during spherical gravitational collapse.

The parameter R_s—known as the Schwarzschild radius—of a body of mass M is defined by

$$R_s = 2GM/c^2.$$

For a body with mass equal to that of the sun, R_s is about 2 miles. When the gravitational collapse of a body begins, its radius R is larger than R_s—in fact, usually much larger. However, as the collapse of the body proceeds, R decreases rapidly and soon becomes equal to R_s. At this point, the outer surface of the body becomes what is called a *trapped surface*. By definition, a trapped surface is a two-dimensional surface (for example, a sphere) on which the gravitational field is so strong that light rays directed outward from this surface "fall back inward." (By "fall back inward" we mean that the surface area of the wave front of light rays decreases rather than increases. In the case of spherical symmetry, that is equivalent to saying that the radius of the wave front decreases rather than increases with time. This is an extreme example of the "light-bending" effect mentioned in 3.5.) After this stage, nothing further that happens to the collapsing body can affect the region of spacetime outside the Schwarzschild radius. In particular, any light ray or particle emitted from the body will never reach the region of spacetime where $R > R_s$; rather it will be pulled back toward the center of spherical symmetry by the strong gravitational field.

To an observer who has the misfortune of being present on the surface of the body as the gravitational collapse proceeds, nothing special would appear to happen when $R = R_s$. The density of matter and the curvature of spacetime are both still finite. However, the collapse continues beyond this point at a rapid rate. Within a short time—of the order of only 10^{-5} seconds for a solar mass body—the radius of the body goes to zero and an infinite density, infinite spacetime curvature singularity, somewhat reminiscent of the big bang singularity, is created.

An observer who stays safely away from the collapsing body never sees the creation of this singularity—or, for that matter, anything that happens to the body after it crosses its Schwarzschild radius R_s—for the simple reason that no light or anything else emitted from the region inside R_s can reach him. Since the light that was emitted by the body just before it crossed R_s takes progressively longer and longer to reach this outside observer,[1] in principle he would see the collapse of the body slow down and the body asymptotically approach its Schwarzschild radius. It should be emphasized that this apparent behavior is merely an optical effect, not a true description of what is physically happening to the collapsing body. In practice, the apparent luminosity of the body will decrease very rapidly in time and the light will also be greatly red shifted, so that rather than see the body asymptotically approach R_s, very soon the outside observer will see no light at all coming from the body. The region of spacetime within the Schwarzschild radius R_s will appear completely black.

Our outside observer could, if he wishes, go to this black-appearing region and determine for himself what is happening there. If he begins his journey too late, he can never catch up with the collapsing body; it will have disappeared into the infinite curvature spacetime singularity before he arrives. However, the gravitational field of this collapsing body persists forever. In particular, if our observer is foolish enough to cross into the region $R < R_s$, he will never

---

1. In general relativity as in special relativity, any observer making a local measurement of the speed of light in vacuum will always obtain the result c. It is the curved spacetime geometry in the vicinity of the collapsing body, not a variation of the speed of light, which is responsible for making the light emitted from the body take "longer and longer" to reach an outside observer.

be able to escape from this region. No matter how powerful a rocket ship he may have, he will get pulled into the central singularity within a finite time after crossing the Schwarzschild radius.

Because of the properties described in the last two paragraphs, we call the region of spacetime with $R < R_s$ a *black hole*. The particular type of black hole formed by spherical collapse is called a Schwarzschild black hole after the man who first discovered the solution (see 7.3). A general definition of the term *black hole* may be given as follows: A spacetime is said to be *asymptotically flat* if the curvature of spacetime approaches zero (that is, the gravitational field becomes weaker and weaker) at large distances from some central region in the spacetime. Thus, the spacetime geometry (gravitational field) associated with an isolated body (such as the body undergoing collapse which we have been considering) is asymptotically flat. A *black hole* in such a spacetime is a region of the spacetime having the property that the gravitational field is so strong that anything which enters this region can never escape from it. The boundary of a black hole is called the *event horizon*. In the spherical case, the event horizon is located at the Schwarzschild radius R_s. By definition, once one crosses the event horizon and enters a black hole, one can never again go back to the distant part of the spacetime where the gravitational field is weak. On the other hand, an observer who remains outside the black hole can never see anything that takes place inside the black hole.

One often is asked whether the universe itself—specifically the closed universe model which recollapses (see chap. 4 and 5)— might be a black hole. Indeed there is a close analogy between the behavior of a body which undergoes complete gravitational collapse and the collapse of the entire universe; both result in infinite density and infinite spacetime curvature in a finite time. However, the whole notion of a black hole requires the existence of regions of spacetime outside the black hole, that is, the asymptotically flat region mentioned above; otherwise the statement that "nothing can escape" is trivial. Thus, the concept of a black hole cannot meaningfully be applied to the universe as a whole.

Figure 21 is a spacetime diagram illustrating the gravitational collapse of a spherical body. (Note that in drawing spacetime diagrams in general relativity we not only have the problem of repre-

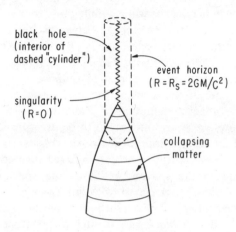

black hole
(interior of
dashed "cylinder")

event horizon
$(R = R_S = 2GM/C^2)$

singularity
$(R = 0)$

collapsing
matter

Figure 21   A spacetime diagram of spherical gravitational collapse (see text).

senting a four-dimensional spacetime on a two-dimensional sheet of paper, but we also have the problem of representing a *curved* geometry on *flat* paper.) The (nearly) horizontal lines near the bottom of the diagram represent the spherical surface of the body at various stages of collapse. As the collapse proceeds in time (that is, going upward in the diagram) the radius of the star decreases and eventually becomes zero; a spacetime singularity (represented by the jagged line) results. The surface indicated by the dashed line is the event horizon; the region inside this surface is a black hole. Any particle or light ray which enters the black hole gets pulled into the spacetime singularity at the center of the black hole.

7.2   The Cosmic Censor Hypothesis

The important features of spherical gravitational collapse have been outlined above. However, no realistic gravitational collapse will ever be exactly spherical; the collapsing body will surely be at least slightly deformed and in many cases it will not even be close to spherical. What happens in a general, nonspherical gravitational collapse?

As mentioned at the beginning of the previous section, a direct analysis of nonspherical gravitational collapse has thus far proven to be intractable. Thus, one must resort to general theorems and, unfortunately, guesses as to what must happen.

The first question one might ask is, Will a spacetime singularity always be produced by gravitational collapse? The answer to this question is yes. In the singularity theorem given in 4.5, hypothesis 3 may be replaced by the requirement that the spacetime possess a trapped surface (defined in 7.1). At least in the case where the gravitational collapse does not differ too greatly from spherical symmetry, one can show that a trapped surface must be present. Thus, a spacetime singularity must occur. Again, the precise nature of the singularity is not specified by the theorem. We shall comment further on this in 7.5.

Granted that a spacetime singularity is produced, there remain two possibilities: (1) either the singularity is contained within a black hole region of the spacetime as in the spherical case, *or* (2) it is possible, in principle, for an observer to get as close to the spacetime singularity as he wishes and still escape out to large distances from the singularity (where he can tell his friends about it). If the latter alternative holds, the singularity is said to be "naked." Does gravitational collapse always result in a black hole, or can naked singularities sometimes be produced? The answer to this extremely important question is not yet known for certain; this is an active area of current research in general relativity. However, all available theoretical evidence points to the validity of the following conjecture, which is quite aptly called the "cosmic censor hypothesis."

Cosmic Censor Hypothesis: No naked singularities occur. In other words, the complete gravitational collapse of a body always produces a black hole; all spacetime singularities resulting from collapse are contained within the black hole and are thus hidden from the view of outside observers.

We shall discuss some theoretical evidence behind this conjecture in 7.4. Here we shall merely make two comments. First, on the basis of what we have said thus far, this conjecture should seem very reasonable. It is known to be true in the case of spherical symmetry. Why should the situation change drastically in the nonspherical case?

Second, if the conjecture is false, it would make life very unpleasant for physicists. We presently do not have a theory of spacetime singularities which allows us to predict what (if anything) may come out of one. Furthermore, as discussed at the end of 4.5, the theory of general relativity itself presumably breaks down in the extreme conditions near a spacetime singularity. Thus, if a naked singularity were produced, we would lose our ability to predict what will happen to us in the future. On the other hand, if the singularity is contained in a black hole, we still may not know what is happening near it, but we do know that whatever happens in the immediate vicinity of the singularity cannot affect us if we remain outside the black hole. Thus, if the cosmic censor hypothesis is valid, these uncertainties in our understanding of the laws of nature are confined to the deep interior of the black hole; ordinary physics applies outside black holes.

7.3 Black Hole Uniqueness

Let us assume the validity of the cosmic censor hypothesis, at least for the time being. Thus, we assume that gravitational collapse always produces a black hole. What are the possible types of black holes that can result from collapse?

In the first place, one would expect that all matter which is in the vicinity of the black hole will quickly fall into it. Thus, to a good approximation the region outside a black hole should be pure vacuum, except possibly for the presence of electromagnetic fields. (This will not be the case if, say, the black hole is in a close binary orbit with an ordinary star; see 9.5. However, even here the modification to the spacetime geometry caused by this additional matter is negligible.) Second, we expect that following the gravitational collapse, the spacetime geometry outside the black hole will quickly settle down to a *stationary* final state. The word *stationary* means time-independent; by the statement that the geometry (that is, gravitational field) outside the black hole is stationary, we mean that there exists a family of observers outside the black hole such that to them everything "looks the same" as time proceeds. We believe that a black hole must become stationary in this sense because if it did not, then presumably energy in the form of gravitational radiation (see 9.4) would continue to be radiated away forever, which seems highly implausible.

Taking the above remarks into account, the question of the possible final states of gravitational collapse translates into the following mathematical question, What are all the stationary solutions of Einstein's equation with no matter present (except possibly for electromagnetic fields) which represent spacetimes with a black hole?

The first class of such solutions was discovered by Karl Schwarzschild in 1915, only a few months after Einstein's formulation of general relativity. These solutions describe the type of black hole formed by spherically symmetric gravitational collapse (7.1). (Of course, their interpretation as black holes came long after Schwarzschild's discovery.) The Schwarzschild solutions are a one-parameter family of solutions: for each value of the mass M of a black hole, there is one solution in the Schwarzschild class.

Within several years of Schwarzschild's discovery, H. Reissner and G. Nordström found a generalization of Schwarzschild's solutions which describe a black hole possessing both mass M and electric charge Q. When $Q = 0$ these solutions reduce to Schwarzschild's. Astrophysical bodies would not be expected to have a net electric charge, so these solutions are valuable much more for their mathematical interest rather than their astrophysical importance.

No further solutions were discovered until 1963, when Roy Kerr discovered a family of solutions describing black holes possessing mass M and angular momentum J. Such a black hole presumably would be formed by the collapse of a spinning body. Again, when $J = 0$ Kerr's solutions reduce to Schwarzschild's. Finally, several years after Kerr's discovery, E.T. Newman and co-workers generalized both the Reissner-Nordström and Kerr discoveries by finding a three-parameter family of solutions describing a black hole with mass M, angular momentum J, and charge Q. This generalized family is known as the Kerr-Newman family of solutions. It should be mentioned that Kerr's discovery, as well as Newman's generalization, arose from mathematical attempts to find *any* new solutions to Einstein's equation, not specifically solutions describing black holes or other objects of physical interest. The true importance of Kerr's discovery only became evident years after the solutions were found.

The three-parameter family of Kerr-Newman solutions encompass all the known solutions which describe black holes. Furthermore, the

Kerr-Newman black holes must satisfy the relation

$$\left(\frac{GM}{c}\right)^2 \geqq G\left(\frac{Q}{c}\right)^2 + \left(\frac{J}{M}\right)^2.$$

If the inequality is violated, the Kerr-Newman family still yields solutions of Einstein's equation, but they describe naked singularities rather than black holes.

As has already been mentioned several times in this book, it is very difficult to find exact solutions of Einstein's equation. Hence the fact that the Kerr-Newman solutions are the only known black hole solutions is by itself very weak evidence that they are the only such solutions that exist. However, within the past ten years, a remarkable sequence of theorems have been proven which establish that they indeed are the only such solutions.

In 1971, Stephen Hawking proved that a stationary black hole must possess an additional symmetry: it must be either *static* or *axisymmetric* (or both). (*Static* means that in addition to being stationary, everything would also "look the same" if the direction of time were reversed; in particular, *static* implies nonrotating $[J = 0]$. *Axisymmetric* means that "everything would look the same" if one performed an arbitrary rotation about a given axis; for example, any figure of revolution is axisymmetric.) But several years earlier Werner Israel had already proven that the Reissner-Nordström solutions (Kerr-Newman solutions with $J = 0$) are the only possible ones which possess the static symmetry. Finally in 1975 David Robinson, following up on some earlier work of Brandon Carter, proved that the Kerr solutions were the only possible pure vacuum (that is, no electromagnetic fields, $Q = 0$) black holes which possess axisymmetry. He also obtained a slightly weaker form of this result in the case where electromagnetic fields may be present. Taken together, these results allow us to conclude that the Kerr-Newman family of solutions completely describes all the stationary black holes which can possibly occur in general relativity.

7.4   Gedankenexperiments to Test the Cosmic Censor Hypothesis

The cosmic censor hypothesis plus the results described in the previous section yield the following truly remarkable conclusion:

The complete gravitational collapse of a body always produces a Kerr-Newman black hole; the properties of this black hole are completely specified by giving its mass M, angular momentum J, and charge Q. For an uncollapsed body, M, J, and Q are only three of an infinite number of quantities which would be required to specify the size, shape, and other properties of the body. (For example, other properties would include its mass quadrupole moment, its magnetic dipole moment, and so on, as well as the information concerning what type of matter the body is made of.) Suppose we start with two uncollapsed bodies which have the same mass, angular momentum, and charge but differ in all other respects (for example, they are made of completely different materials and have completely different sizes and shapes). Now, let each of these bodies undergo complete gravitational collapse. The above conclusion states that the black holes thereby produced will be completely indistinguishable from each other!

In other branches of physics, when one arrives at such a striking conclusion via reasoning based partly on solid results and partly on pure conjecture, there is a very natural thing to do: go into the laboratory and do an experiment to see if nature is in accord with this conclusion. Clearly this option is not open to us here. We know of no force powerful enough to induce gravitational collapse except gravity itself and, as discussed in chapter 6, this requires bodies with mass greater than that of the sun. Thus, we certainly cannot do a laboratory experiment where we collapse different types of objects and see if they always produce Kerr-Newman black holes.

However, we can do the next best thing: gedankenexperiments, that is, thought experiments. We can consider some simple examples of gravitational collapse where we can calculate directly what happens without relying on conjectures like the cosmic censor hypothesis; then we can see if the result is in accord with the above conclusion. Clearly, a gedankenexperiment is a poor substitute for a real experiment since it does not test the validity of the underlying theory (general relativity), but it is the best we can do. Since the cosmic censor hypothesis is the weakest link in the chain of reasoning leading to the above conclusion, the gedankenexperiments given below are basically tests of the validity of the cosmic censor hypothesis in general relativity.

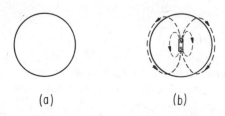

(a)   (b)

Figure 22   (a) A spherical shell of matter (mass M) and (b) a similar spherical shell (mass M) with a magnet placed at its
center. The gravitational collapse of shell (b) produces
a Schwarzschild black hole identical in all its properties to the black hole that results from the collapse of
shell (a).

As a first gedankenexperiment, consider the collapse of a spherical
shell of matter of mass M and the collapse of a similar shell of matter
also of mass M which has an ordinary magnet placed inside the
shell (see fig. 22). As discussed in 7.1, if the first shell undergoes
gravitational collapse, it will produce a Schwarzschild black hole.
According to the above conclusion, since the second shell has the
same mass M, angular momentum $J = 0$, and charge $Q = 0$ as the
first shell, its gravitational collapse must also produce a Schwarzschild black hole identical in all its properties to the first one. How
can this be possible, since a Schwarzschild black hole has no magnetic field, whereas on account of the presence of the magnet one
would expect the second collapsed body to possess a magnetic field?
The calculation of what actually happens in this simple example is
relatively easy, and here are the results.

In the prerelativistic Newtonian theory of gravity, a massive body
(possessing no charges or currents of its own, as we are assuming)
cannot have any influence on a magnetic field. In general relativity
it remains true that a massive body does not affect a magnetic field
in a direct, local manner. However, by curving the spacetime geometry, a massive body can have a very important global influence on
the structure of the magnetic field. In this example, it turns out that
as the radius of the second shell approaches its Schwarzschild radius,
the strength of the magnetic field outside the shell becomes greatly

diminished. The complete gravitational collapse of the shell does indeed produce a Schwarzschild black hole, with no magnetic field remaining. Our conclusion is validated in this example.

As a second gedankenexperiment, let us attempt to destroy a black hole and create a naked singularity. Recall from the previous section that Kerr-Newman black holes satisfy the relation

$$\left(\frac{GM}{c}\right)^2 \geqq G\left(\frac{Q}{c}\right)^2 + \left(\frac{J}{M}\right)^2.$$

The idea is to start with a black hole for which equality holds in the above relation, and then make it capture something the charge and/ or angular momentum of which are large enough so that the above inequality will be violated. If we can do this, the resulting object can no longer be a black hole; the singularity which was inside the black hole will become naked and the cosmic censor hypothesis will be violated.

As a first example, let us start with a black hole with $\sqrt{G}M = Q$ and $J = 0$. If we drop in a charged particle of mass m and charge q satisfying $\sqrt{G}m < q$, we will destroy the black hole. There is, however, one problem with this. If $\sqrt{G}m < q$, the Coulomb electrostatic repulsion on the particle becomes greater than the gravitational attraction. Thus, if we let go of such a particle, it will not fall into the black hole; rather, it will fly away from it! So, this doesn't work. However, we can throw such a particle toward the black hole with great enough speed so that it will go in. But when we do this, we increase the energy of the particle and hence the mass increment it gives to the black hole by just the right amount so that when the black hole captures it the inequality $\sqrt{G}M \geqq Q$ will be maintained. Thus, this attempt to destroy a black hole fails.

As a second example, we start with a black hole satisfying $GM^2/c = J$ and try to make it capture a particle with large enough orbital angular momentum to violate this inequality. Again we find we cannot do this. In order to give the particle the required angular momentum we have to send it in with such a large impact parameter that it misses the black hole entirely. Another possibility is to drop into the black hole a particle which is spinning so rapidly that the angular momentum it would contribute to the black hole would cause violation of the inequality. It might appear that this will work.

However, in general relativity spinning bodies do *not* move on geodesics; there is an additional spin-spin force between two rotating bodies. (This additional force is very similar in character to the force between two magnets in electromagnetism.) Ordinarily this spin-spin force is negligible, but in the vicinity of a "maximally rotating" (that is, $GM^2/c = J$) black hole this force is just strong enough to prevent a particle which is spinning rapidly enough to violate the inequality from entering the black hole. Thus, this attempt also fails.

These gedankenexperiments as well as other results which we have not discussed greatly increase our confidence in the validity of the cosmic censor hypothesis and the key conclusion derived from it: *The complete gravitational collapse of a body always produces a Kerr-Newman black hole.* For the remainder of this book we shall accept this conclusion.

7.5   What Is inside a Black Hole?

What is inside a black hole? The singularity theorem (see 4.5 and 7.2) assures us that some sort of spacetime singularity will be found inside a black hole. For the type of black hole formed by spherical collapse, this spacetime singularity is all-encompassing in the sense that any observer who enters the black hole will get pulled into it. Must this be the case for all types of black holes? The answer is no.

As discussed in the previous sections, we expect the exterior geometry (that is, gravitational field) of a black hole formed by collapse to be described by a Kerr-Newman solution. However, for a realistic gravitational collapse (for example, one that is not precisely spherical) we have no reason to expect the Kerr-Newman solutions to accurately describe the interior geometry of a black hole. Nevertheless, we can analytically continue the exterior Kerr-Newman solution to the interior and see what we get. When we do this for a solution with $J \neq 0$ or $Q \neq 0$ (that is, anything but a Schwarzschild solution) we find behavior that seems like a science fiction writer's idea of a spacetime warp.

In the first place, there is, of course, a spacetime singularity inside the black hole, but not everything that enters the black hole must go into the singularity; in fact, a particle has to be rather carefully aimed in order to fall directly into the singularity. More remarkable than this, however, is the fact that, roughly speaking, there is a truly infinite amount of spacetime contained within the black hole. In principle, an observer who enters the black hole in a rocket ship can actually steer away from the singularity and out to a new asymptotically flat region of spacetime! In other words, while still inside the black hole—he can, of course, never return out of it—he can get far enough away from the singularity that the gravitational field is negligible and he is out of any danger of getting pulled back into it. If he chooses, he can live out the rest of his life normally in this region. But if he is truly adventurous, he will notice that a new black hole (that is, a black hole in this new asymptotically flat region, all contained within the original black hole) has formed around the region where he had previously escaped from the singularity. If he wishes, he can enter this new black hole. Again he can steer away from the singularity and enter yet another new asymptotically flat region of spacetime, where he will encounter yet another new black hole, and so on!

Before the adventurous reader decides to attempt this journey, let me caution him that the best theoretical evidence presently available indicates that the interior of a black hole formed in a realistic gravitational collapse is probably *not* like the Kerr-Newman model described above. Rather, it is probably like the interior of a Schwarzschild black hole, with an all-encompassing spacetime singularity. Even so, the analytically continued Kerr-Newman solutions yield a fascinating example of the type of spacetime structure possible in the theory of general relativity.

Eight

Energy Extraction from Black Holes

A black hole is a region of spacetime where the gravitional field is so strong that nothing—not even light—that enters this region can ever escape from it. Therefore, it would seem that about the last thing possible would be extraction of energy from a black hole. However, it turns out that for a rotating black hole (that is, a black hole with angular momentum $J \neq 0$ formed by the collapse of a spinning body) energy extraction is indeed possible; in fact, in principle a substantial fraction of the mass of a rotating black hole can be converted to useful energy. The ideas which lie behind this rather exotic energy extraction process are the subject of this chapter. We shall begin, however, by giving a (somewhat overdue) discussion of the notions of energy in prerelativity physics, in special relativity, and in general relativity.

8.1 Energy and Momentum in Prerelativity Physics

The concept of energy is at least somewhat familiar to anyone who has ever paid an electricity bill. To each particle and field we associate a number which we call its energy. For example, in prerelativity physics the energy of a free particle is $mv^2/2$, where m is the mass of the particle and v is its velocity; the energy density (that is, energy per unit volume) associated with an electric field of strength $\vec{E}$ is $|\vec{E}|^2/8\pi$. The total energy of a system is defined to be the sum of the energies of all the particles and fields present in the system.

Energy may be transferred from one entity to another, for example, the energy of a particle may be converted to the energy of a field or the energy of another particle. Indeed, we pay the power companies to convert the energy of waterfalls, coal, and atoms to the mechanical energy of our motors and the thermal energy of our light bulbs. However, in all these processes the total energy remains unchanged; total energy is conserved.

In a similar manner, in prerelativity physics we associate to each particle and field a vector called momentum. The momentum of a free particle is $m\vec{v}$. Again, the total momentum of a system is conserved.

8.2   Energy-Momentum in Special Relativity

In special relativity, for the reasons discussed in 3.1, the prerelativity notions of energy and momentum are unsatisfactory and need to be reformulated. What direction in spacetime is the prerelativity momentum vector $m\vec{v}$ of a particle supposed to point in? With our prerelativity notions of space and time (see chap. 1), momentum points in a spatial direction, that is, it is tangent to an absolute "surface" of simultaneity. But in special relativity there are no absolute surfaces of simultaneity, so the old prescription of momentum $= m\vec{v}$ is meaningless as it stands.

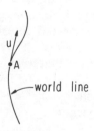

Figure 23   A spacetime diagram showing the world line of a particle and the tangent vector **u** to the world line at event A.

The new notion of energy and momentum in special relativity is as follows: Consider a particle in spacetime. The world line of this particle is illustrated in figure 23. Let **u** denote the tangent vector to this curve in spacetime. We define the energy-momentum "four-vector" **e** of the particle by the formula

$$\mathbf{e} = m\mathbf{u}$$

where m is the mass of the particle. (Some authors refer to m as the "rest mass.")

To break down the concept of energy-momentum **e** into energy and momentum, we need to introduce an observer. Consider an inertial observer O who is present where the particle is located, that is, his world line intersects the world line of the particle. Let **t** denote the tangent vector to the observer's world line. At the event of the intersection of their world lines, the energy E of the particle (as measured by O) is defined to be the component of **e** along the direction **t**. (In other words, it is the scalar product of **e** and **t**, where the scalar product is taken using the metric of spacetime.) The momentum of the particle (as measured by O) is defined as the part of the vector **e** which lies in the surface of (relative) simultaneity determined by O. The energy and momentum of fields can be defined in a similar manner. With each field one associates an object called the "stress-energy-momentum tensor," components of which give the energy density and momentum density of the field.

If observer O is not present where the particle is located, there still is no difficulty in defining energy in special relativity. In this case, the energy of the particle "as measured by O" is defined as the energy measured by an inertial observer who is present at the site of the particle and who moves with zero velocity relative to O. Another way of saying this is that in all cases, the energy of a particle as measured by O is defined to be the time component of **e** in the global inertial coordinate system (see 2.1) determined by O. Similar remarks apply for momentum.

The total energy of a system as measured by observer O is defined as the sum of all particle and field energies as measured by O. Total momentum is defined in a similar manner. Again, total energy and momentum are conserved.

The new notion of energy in special relativity has one extremely important implication. In prerelativity physics, one could add an arbitrary constant to the formula for the energy of a particle without affecting anything. For convenience, we chose this constant so that the energy of a particle at rest is zero. But with our new notion of energy in special relativity, we no longer have the freedom to add an arbitrary constant in the formula for energy. In fact, the energy,

E, of a particle as measured by an observer who is at rest with respect to the particle is given by the famous formula

$$E = mc^2.$$

The fact that there is no arbitrariness in this value of E for a particle at rest suggests that we should take it seriously. It suggests that rest mass should be viewed as a form of energy, and it opens up the possibility of converting energy in the form of mass to other forms of energy in accordance with the above formula. This, of course, is one of the most striking predictions to arise from the theory of special relativity.

8.3   Energy in General Relativity

In general relativity, the energy momentum "four-vector" **e** of a particle and the "stress-energy-momentum tensor" of fields are defined in the same manner as in special relativity. Indeed it is really the full stress-energy-momentum tensor of matter (rather than just the energy density) which is set equal to a spacetime curvature tensor in Einstein's equation (see 3.4). However, there is one major exception to the association of stress-energy-momentum tensors with fields: the gravitational field itself is described purely in terms of the curvature of spacetime and it does not have any quantity associated with it that describes its local energy and momentum density.

An observer can define the energy of a particle in his immediate vicinity in exactly the same manner as described in the third paragraph of 8.2. However, in general relativity there is no such concept as a "global inertial coordinate system" or a "distant inertial observer who remains at rest with respect to a given inertial observer." Hence, an observer can only define the energy of matter in his vicinity. There is no analogue of the procedure described in the fourth paragraph of 8.2.

Thus, with a notable exception to be discussed below, there is no notion of total energy in general relativity for two reasons: (1) There is no notion of the energy of distant matter, and (2) there is no notion of gravitational field energy. Thus, there is no meaning to the phrase "the total energy content of the universe" and one cannot even ask about conservation of total energy. However, in general relativity one still has conservation of energy in the following local sense: in any small region around a freely falling observer (where *small* means small compared with the lengths associated with the curvature of spacetime) the (nongravitational) energy increase (or decrease) in this region must equal the flux of energy through the boundary of the region.

The exception mentioned above where a definition of total energy is possible is the case of an asymptotically flat spacetime. As defined in 7.1, as asymptotically flat spacetime is one in which the space-time curvature (that is, gravitational field) becomes smaller and smaller as one gets far away from the central region of the spacetime; in other words, it describes the gravitational field of an isolated system (for example, a star or black hole). As one gets far away from the central region in an asymptotically flat spacetime, the gravitational field appears more and more like an ordinary Newtonian gravitational field, that is, the effects of general relativity are truly significant only in the very strong field region near the central body. Therefore, one can define the *total mass M* of the system as the value of the mass that would be required in the Newtonian theory of gravity to produce the same gravitational field at large distances. Following the lesson learned in special relativity, we define the *total energy* ϵ of the system by the formula $\epsilon = Mc^2$. One can show that the total energy is conserved, that is, its value cannot change with time.

In general, there is no known direct relationship between the total energy ϵ and the local energy of the matter in the spacetime. However, if the spacetime—in addition to being asymptotically flat —is also stationary, then there is a formula directly expressing ϵ in terms of the local energy of matter. Recall that a stationary space-time is one in which the spacetime geometry does not change with time. This time translation symmetry defines a preferred direction at every point in spacetime which we shall denote by **t**. To an observer whose world line is everywhere tangent to this direction **t** (that is, to a stationary observer), everything would "look the same" as time proceeds. In a stationary spacetime each particle contributes an energy E to the total energy of the spacetime given by the compo-nent of its energy momentum "four-vector" **e** along the direction **t**.

One may think of E as the sum of the rest mass, kinetic energy, and gravitational potential energy of the particle. For a freely falling (geodesic) particle, the particle energy E is constant in time. Thus, in the case of a stationary, asymptotically flat spacetime, one recovers much of the notion of energy that exists in special relativity.

8.4   t Spacelike? The Ergosphere

Recall that timelike curves represent all the possible motions in spacetime of observers. (In particular, timelike geodesics represent the motions of all freely falling observers.) A vector is said to be *timelike* if it is tangent to a timelike curve. On the other hand, a vector is said to be *spacelike* if it is tangent to a curve in spacetime having the property that nearby points on this curve are spacelike separated. What, then, does it mean to say that in a stationary spacetime, the *time* translation symmetry vector **t** is *spacelike* in some region?

Recall that **t** is the direction in spacetime in which one would have to travel in order for "everything to look the same" as time goes on and also in order that one would appear to be "standing still" according to other stationary observers. Roughly speaking, if **t** is spacelike in a region, it means that in order to "stand still" in this region, one would have to travel on a spacelike curve, that is, go faster than the speed of light! In other words it means that in this region it is impossible to "stand still" relative to stationary observers outside this region.

Where might one expect to find a region where **t** is spacelike? Inside a black hole. For example, outside a Schwarzschild black hole the spacetime geometry is stationary; with a powerful enough rocket ship one can stand still at any radius outside the black hole. But once one enters the black hole it is impossible to stand still; one gets pulled into the central spacetime singularity. Thus, inside a Schwarzschild black hole **t** is spacelike.

It is a remarkable fact, however, that there is also a region outside a rotating black hole (that is, a black hole with $J \neq 0$, formed by the collapse of spinning matter) where **t** becomes spacelike. This region, called the *ergosphere,* is schematically sketched in figure 24. Why does **t** become spacelike outside a rotating black hole? One can understand why in terms of Mach's principle (see 3.2). The

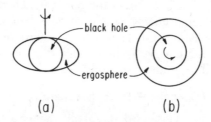

Figure 24   A schematic representation (*not* a spacetime diagram) of the ergosphere of a rotating black hole, showing (*a*) a side view and (*b*) a top view.

rotating matter which collapsed to form the black hole has a strong influence on the motion of the bodies near the black hole. Bodies which come close enough to the black hole so that they enter the ergosphere are forced to rotate in the same direction as the matter which collapsed to form the black hole. Thus, in the ergosphere, even though it is possible to escape from the strong gravitational field of the black hole, it is impossible to "stand still," that is, **t** is spacelike.

The fact that **t** can become spacelike has a very important consequence. The energy-momentum vector **e** of a particle must always be a timelike vector. If **t** is also timelike, then one can show that the component of **e** in the direction of **t** must be positive. In other words, the contribution E that the particle makes to the total energy ϵ must be positive. But if **t** is spacelike, then E can take either sign (or be zero). Thus, in the ergosphere, there can exist particles which have negative energy E. As described in the next section, we can use these particles to extract energy from the black hole.

8.5   Energy Extraction from a Rotating Black Hole: The Penrose Process

We now shall describe how to extract energy from a rotating black hole by a method proposed by Roger Penrose in 1969. Start with a rotating black hole of mass M (energy Mc^2) and, far from the black hole, a particle of energy E_0. Since the particle is outside the ergosphere, E_0 must be positive. Now, let the particle fall into the ergosphere. When it reaches the ergosphere arrange for it (say, by the use of explosives and a timing device) to break apart into two fragments in such a way that the first fragment has *negative* energy E_1 and falls into the black hole while the second fragment escapes back out to large distances from the black hole. (The energy of the explosive is, of course, included in E_0.) This process is illustrated in figure 25.

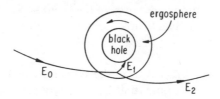

Figure 25   The Penrose energy extraction process (see text).

By local conversation of energy-momentum we have

$$E_0 = E_1 + E_2.$$

Since E_1 is negative, the energy E_2 of the fragment that escapes is *greater* than the incident particle energy E_0. Furthermore, when the first fragment goes into the black hole, it *reduces* the black hole's mass-energy. Thus, at the end of this process we are left with a black hole of energy $(Mc^2 - |E_1|)$ and a particle of energy $(E_0 + |E_1|)$. In other words, we have extracted the energy $|E_1|$ from the black hole. The trick which enabled us to perform this feat was to make the black hole swallow negative energy.

8.6   Energy Extraction Limits

Is there any limit to the amount of energy which can be extracted from a rotating black hole by repeated application of the Penrose process? Yes. It turns out that a negative energy particle in the ergosphere must also have negative orbital angular momentum, that is, angular momentum opposite that of the black hole. Thus, in the Penrose process, when the negative energy particle enters the black hole, it also reduces the angular momentum of the black hole. If one were to repeat the Penrose process indefinitely, eventually the angular momentum of the black hole would be reduced to zero ($J = 0$). When this happens, there is no longer an ergosphere and no more energy can be extracted.

How much energy can be extracted from a rotating black hole? One could explicitly calculate the result of repeated application of the Penrose process until J is reduced to zero. However, there is a simple, yet very powerful, method by which one can obtain energy extraction limits. This method relies on two facts. The first is a theorem concerning the surface area of the event horizon (that is, the outer boundary) of a black hole, which was proven by Stephen Hawking in 1971.

Area Theorem: The surface area of the event horizon of a black hole can never decrease with time, that is, after any process the surface area of a black hole must always be larger than (or equal to) its initial area.

The second fact on which the method relies is the formula for the horizon area of a Kerr black hole. (We shall consider only the uncharged case, $Q = 0$, here, although the generalization to $Q \neq 0$ is not difficult.) The horizon area, A, can be calculated directly from the Kerr spacetime metric and, of course, it (as well as everything else) depends only on M and J. The formula is

$$A = \frac{8\pi G^2 M^2}{c^4} (1 + \sqrt{1 - (cJ/GM^2)^2}).$$

For fixed M, the area A is largest for a Schwarzschild black hole ($J = 0$; $A = 16\pi G^2 M^2/c^4$) and smallest for a maximally rotating Kerr black hole ($J = GM^2/c$; $A = 8\pi G^2 M^2/c^4$).

Suppose we start with a maximally rotating Kerr black hole of mass M_i and perform some energy extraction process like the Penrose process. How much energy can be extracted? We know the initial area of the black hole is $A_i = 8\pi G^2 M_i^2/c^4$. After completion of the process, the black hole will again "settle down" to a stationary Kerr state and by the above equation its final area A_f will be related to its final mass, M_f, by the inequality $A_f \leq 16\pi G^2 M_f^2/c^4$. But by the area theorem we have $A_f \geq A_i$. Putting these inequalities together, we obtain $M_f^2 \geq \frac{1}{2} M_i^2$. This tells us that at most $(1 - 1/\sqrt{2}) = 29\%$ of the original mass-energy $M_i c^2$ can be extracted. One can think of this extractable energy as the rotational energy of the original black hole.

The figure 29% represents an extremely high fraction of conversion of mass to other forms of energy. The most energy-releasing process we can perform in a laboratory—the fusion reaction converting hydrogen to helium—converts less than 1% of the original mass of the hydrogen to energy. Thus, rotating black holes are actually the most efficient energy sources we know of. However, it does not appear that black holes will provide a solution to the energy crisis brought about by our eventual exhaustion of fossil fuel. Even if we should find a rotating black hole in our vicinity, the Penrose energy extraction process is not very practical since it requires a very large break-up velocity of the particle into fragments, as well as extremely accurate aim and timing. Thus, the energy extraction ideas discussed here—as well as all the other ideas in this book—are (at least at the present time) of interest for understanding nature rather than for possible practical, technological advances.

The method for obtaining energy extraction limits can be applied to even more exotic processes, such as "colliding" black holes. Suppose we have two Schwarzschild ($J = 0$) black holes present initially, with masses M_1 and M_2. Like two ordinary massive bodies, these black holes will be pulled toward each other by their mutual gravitational attraction. Presumably they will eventually "collide" and coalesce, producing a single, large Schwarzschild black hole of mass M_3; gravitational radiation (see 9.4) will also be emitted during this process. I say "presumably" because the calculation of what actually happens is so difficult that only preliminary results of numerical calculations using computers are presently available. However, in a very simple manner, we can obtain an upper limit to the amount of energy that can be carried off in the form of gravitational radiation.

First we have the formula for the area of the initial and final black holes in terms of their masses, $A = 16\pi G^2 M^2/c^4$. Second, the area theorem tells us that $A_1 + A_2 \leqq A_3$, where A_1 and A_2 are the areas of the two initial black holes, and A_3 is the area of the final black hole. Finally, we know that by conservation of total energy, the energy carried off by gravitational radiation, E_r, is given by

$E_r = M_1c^2 + M_2c^2 - M_3c^2$. Putting these three facts together, it is not difficult to find an upper limit on E_r. (Calculations indicate that the actual amount of energy radiated away in this process will be considerably less than the upper limit given by this method.)

Nine

The Astrophysics of Black Holes

In this chapter, we shall focus on the effects that might be produced by the presence of a black hole in a realistic astrophysical environment. Our principal concern will be, How might we observe—or, more precisely, deduce the presence of—a black hole? This is a very important question, since the observation of a black hole would provide strong confirmation of the basic ideas underlying general relativity and would give us a testing ground for predictions of the theory in regions of strong gravitational field.

9.1 Formation of Black Holes; Primordial Black Holes

How are black holes formed? There are three distinct processes which conceivably can lead to gravitational collapse and formation of a black hole.

1. *Gravitational collapse of a star:* As described in chapters 6 and 7, for sufficiently massive stars the natural course of stellar evolution results in gravitational collapse to a black hole. Black holes produced in this manner should have masses ranging between 1 and 50 times the mass of the sun. (Stars with mass below this range can support themselves by electron or neutron degeneracy pressure [see chap. 6] and will not collapse; stars with mass much greater than about 50 solar masses probably do not exist because of pulsational instability.) At present, we can only make crude estimates as to how many black holes should exist in our galaxy as a result of this process. Supernovas are estimated to occur in our galaxy at the rate of about one or two per century. If a high fraction of them result in the formation of a black hole, there could be as many as one hundred million (10^8) black holes in our galaxy.

2. *Collapse of a star cluster:* This possible means of black hole formation is somewhat more speculative than possibility 1. The

central region of a galaxy is densely populated with stars. Elsewhere within galaxies, there are also many dense clusters of stars. As time progresses, the dynamic evolution of a cluster of stars causes the central region of the cluster to become more and more condensed. At the center of the galaxy or in very dense clusters, this evolution may reach a stage where collisions between the stars become very important, resulting in the disruption of the stars and the formation of a single "supermassive" body. This body may then undergo gravitational collapse to a black hole. If this process occurs at the center of a galaxy, one would expect it to produce a black hole with mass between one hundred thousand (10^5) and one billion (10^9) solar masses; a somewhat smaller black hole might be produced at the center of a dense star cluster.

3. *Primordial black holes:* This possibility is considerably more speculative than the first two. In the very early universe, the density of matter was extremely high (see chap. 4 and 5). It is possible that inhomogeneities in the density at this stage could have resulted in the formation of black holes. In other words, if the matter density was enhanced in some region, then rather than expand with the rest of the universe, gravitational collapse of the matter in this region to form a black hole might have occurred. Such primordial black holes could have practically any mass, depending on the nature of the density inhomogeneities. In particular, very small black holes—with mass much less than that of the sun—could have been produced in this manner. Formation of such black holes in the present universe is not possible since processes 1 and 2 above cannot produce black holes of very low mass.

A very low mass primordial black hole—because of its extremely small size—probably would not accrete much matter during the evolution of the universe and thus would remain small in the present era. If such a black hole were to strike the earth, it would pass right through, absorbing a negligible amount of matter. However, its strong gravitational field would cause a shock wave in the atmosphere. This has led to speculation—reported widely in the news media—that a small black hole might have been responsible for the Tunguska "meteor" event in Siberia in 1908, where trees were knocked down over an area covering many square miles but no meteor crater was found. While this explanation may be as good as many others that have been proposed, it is quite implausible. Even if small primordial black holes were produced, it is extremely unlikely that there are enough of them for there to be any reasonable chance for one to strike the earth.

We shall not discuss the possible observational consequences of small primordial black holes any further in this chapter. However, in chapter 10 we shall find that for very tiny black holes (Schwarzschild radius comparable to the size of an atomic nucleus; mass of about 10^{15} gm, that is, a billion tons) quantum effects become extremely important and lead to important observational consequences. As described in chapter 10, if such primordial black holes were created, these quantum effects would lead to the "evaporation" of the black hole, with a resultant burst of high-energy particles.

9.2   Isolated Black Holes: The Gravitational Lens Effect

Is it possible that one could observe an isolated black hole, that is, a black hole not surrounded by other matter (9.3) or in close binary orbit with a star (9.5)? It might seem that the answer must be no. After all, the Schwarzschild radius of a solar mass black hole is only about 2 miles. So, if our sun were a black hole, even our best telescopes could not quite resolve it, and it's black anyway. If a black hole were close enough to us to be "seen" directly, its gravitational effect on us would be greater than that of the sun.

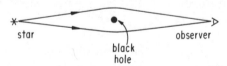

Figure 26   The gravitational lens effect. The "eclipse" of a distant star by a black hole results in enhancement of the light seen by an observer and image distortion due to the "light-bending" effect.

However, it is indeed possible, in principle, to observe a distant, isolated black hole. As illustrated in figure 26, if there is a near "eclipse" of a luminous body by a black hole, the "light-bending" effect (see 3.5) will result in an *enhancement* of the light received as well as distortion of the image of the luminous body. In other words, the black hole acts as a gravitational lens. The amount of light absorbed by the black hole is negligible because the horizon is so small, but light which stays even well outside the horizon will be significantly "bent." Unfortunately, other dense, underluminous objects (for example, a dim white dwarf star) will produce the same gravitational lens effect, so observation of the lens effect would not conclusively establish the presence of a black hole.

If the universe were filled with enough black holes of mass greater than 100,000 solar masses to make the universe closed (see 5.2), then eclipses of quasars by these black holes should occur frequently enough to be observable. However, no such eclipses have been observed. Thus, at present we can only use the gravitational lens effect to set an upper limit on the number of massive black holes in the universe.

9.3   Black Holes in an Interstellar Medium or Star Cluster

A black hole in our galaxy will never be strictly isolated in the sense that no matter whatsoever is in its vicinity. Our galaxy is pervaded by a very tenuous gas with an average density of perhaps one hydrogen atom per cubic centimeter. Thus, a black hole—formed, for example, by the collapse of a star—will be surrounded by this interstellar medium. The interstellar gas near the black hole will fall toward it, and as it does so it will heat up. Like any other hot object it will radiate electromagnetic radiation, probably mainly in the infrared, visible, or ultraviolet range in the case of accretion onto a black hole formed by stellar collapse. Some of this radiation will, of course, go into the black hole, but most of the radiation emitted from outside the black hole will escape to large distances. In principle, this radiation could provide a means of detecting such a black hole. However, the luminosity of this gas probably will be so small that observation of it would be extremely difficult. Thus, it does not appear that this effect will result in the detection of a black hole.

A large black hole at the center of a star cluster (possibility 2 of 9.1) could cause two possibly observable effects: (1) The strong gravitational field of the black hole will affect the distribution of stars in the cluster, making it more densely populated with stars in the region outside the black hole at the center of the cluster. Hence, one could look for a brightness *enhancement* at the center of a star cluster as evidence for the presence of a black hole. Such an effect has not yet been observed. (2) Accretion of matter onto a large black hole could also result in the emission of radiation as discussed above. Very recently, X rays have been observed to come from a number of globular star clusters. It is possible that this X-ray emission could be caused by the presence of a black hole at the center of these clusters.

9.4   Gravitational Radiation

In electromagnetic theory, the acceleration of a charged particle produces electromagnetic radiation (that is, light, radio waves, X rays, and so on). One can think of electromagnetic radiation as "ripples" in the electric and magnetic fields which propagate through spacetime at the speed c.

In general relativity in a nearly flat region of spacetime one finds closely analogous behavior. Acceleration of a mass produces gravitational radiation which can be thought of as ripples in the spacetime curvature, which also propagate at the speed of light c. In a strongly curved region of spacetime the distinction between these ripples and the background curvature of spacetime is unclear and one cannot give a precise definition of gravitational radiation. However, outside a strongly curved region (for example, far away from a black hole) the notion of gravitational radiation is unambiguous. Astrophysical events which are likely to produce large amounts of gravitational radiation are (1) supernovas or other gravitational collapse phenomena (as discussed in 6.4, a supernova also should produce a large burst of neutrinos), (2) accretion of a star into a large black hole at the center of a galaxy or star cluster, (3) collisions of black holes. Hence, the detection of a burst of gravitational radiation will give us information about phenomena possibly involving a black hole. "Gravitational wave astronomy" may well provide a means for detecting black holes.

How may gravitational radiation be detected? If a gravitational wave passes through matter, the ripples in the spacetime curvature will induce stresses in the matter. If these extremely tiny stresses can be measured, one can detect gravitational waves. Work on detection of gravitational radiation began with the pioneering efforts of Joseph

Weber in the 1960s. Although there has been much controversy in this area, it now seems generally agreed that gravitational waves have not yet been detected. One should not be too discouraged by this since the presently available gravitational wave detectors are probably not sensitive enough to detect the levels of gravitational radiation that one would expect from astrophysical sources. However, the gravitational wave detectors which are currently being designed and built may be sensitive enough to detect supernova events in our galaxy and in nearby galaxies.

9.5   Binary Systems Containing a Black Hole: Cygnus X-1

In astronomy, the term *binary system* refers to two bodies (usually stars) which orbit around each other under their mutual gravitational attraction. Binary star systems are very commonly observed throughout our galaxy. Since 1970, when satellites which could detect X rays were first orbited, about ten X-ray sources have been identified with binary systems. In each of these systems one has an apparently normal star in a close orbit with another body which is not seen optically. (The existence of the second body is inferred from the periodic Doppler shift of the spectral lines of the visible star, which shows that it is in orbit.)

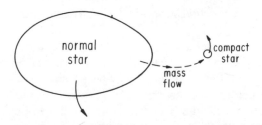

Figure 27   A binary system consisting of a normal star and a compact star (white dwarf, neutron star, or black hole), orbiting around each other in the direction indicated by the arrows. The shape of the normal star is greatly distorted by the tidal influence of the compact star. Mass flow from the normal star to the compact star can result in the production of X rays.

How can X rays be produced by such a binary system? By far the best explanation that has been advanced thus far is as follows: If one has an ordinary star in a close orbit with a compact star (that is, a white dwarf, neutron star, or black hole) one can get mass flow from the normal star to the compact star (see fig. 27). In some cases a steady stream of gas from the normal star can slowly spiral onto the compact star, thereby forming a structure known as the *accretion disk*. Compression and turbulence cause heating of the gas in the accretion disk as it spirals in. This can result in the emission of X rays. In this model it is necessary that the gas accrete onto a compact star (rather than another normal star) in order that the heating be great enough to produce X rays. Also, rapid time variations in the X-ray emission have been observed, indicating that the region where the X rays are emitted probably is very small. Thus, it is generally accepted that all of the binary X-ray sources consist of a normal star orbiting around a white dwarf, neutron star, or black hole.

The binary X-ray source which has received the greatest attention as a candidate for a black hole is Cygnus X-1. From observations of the orbital period and orbital velocity (inferred from Doppler shifts) of the visible star as well as other, more indirect data one can infer the mass of the unseen compact object. The best estimates indicate that the mass of the unseen member of the Cygnus X-1 binary is probably at least 9 solar masses. This is significantly greater than the maximum possible mass of a white dwarf or neutron star (see chap. 6). *Thus, it is concluded that the binary system Cygnus X-1 contains a black hole.*

There are several other binary X-ray sources which are similar in character to Cygnus X-1. However, in these cases the mass of the unseen compact star is smaller and the possibility that it is a white dwarf or neutron star cannot be ruled out. The other binary X-ray sources show periodic X-ray emission similar to pulsars (see 6.5); the compact object in each of these sources is believed to be a neutron star.

It is hoped that further observations in the near future will not only confirm the identification of the compact star in Cygnus X-1 (and perhaps in other binary X-ray sources) as a black hole but will begin to yield detailed information on the properties of the black hole. Observational astronomy relevant to black holes is being actively pursued at present, and one should not be surprised if new and exciting discoveries of things not yet anticipated by theorists are soon made.

Ten

Quantum Particle Creation near Black Holes

Many readers may feel that the properties of a black hole discussed in the previous chapters already make it the most exotic object one would wish to consider. It may seem strange, therefore, that in recent years theorists have begun to consider much more exotic quantum mechanical processes occurring near small black holes. These processes cause the spontaneous creation of particles near a black hole and, in principle, lead to the eventual "evaporation" of the black hole. In this chapter, I will attempt to explain why theorists were led to consideration of these ideas, what the results are, and where this investigation may eventually lead. I begin by describing a well-known process which occurs in atomic physics.

10.1 Stimulated Emission in Atomic Physics

Suppose an atom has two energy levels with energies E_1 and E_2. If an electron occupies the level with the higher energy (E_1) and the lower energy level (E_2) is not occupied, then in general it will be possible for the electron to make a spontaneous quantum mechanical transition to the lower energy state. When it does so, it will emit a photon (that is, a quantum of electromagnetic radiation) with frequency ν given by $h\nu = \Delta E = E_1 - E_2$ where h is Planck's constant. This process is known as *spontaneous emission*.

On the other hand, if one sends in electromagnetic radiation with frequency $h\nu = \Delta E$, then, in general, transitions between the atomic levels will be induced. Suppose one has a gas consisting of many of these atoms. Under ordinary circumstances, in most atoms the electron will be in the lower energy state. The electromagnetic radiation will primarily induce transitions from the lower to the upper level, and one will get *absorption*, that is, the intensity of the electromagnetic radiation will be diminished as it passes through the gas.

However, it is sometimes possible to prepare the gas so that initially in most atoms the electron populates the higher energy level. (Such atoms are said to be in an *excited state.*) In this case, the electromagnetic radiation primarily will induce transitions from the upper to the lower level, resulting in *stimulated emission:* the transmitted electromagnetic radiation will have *greater* intensity than the incident radiation. This is precisely the mechanism by which a laser works.

General arguments (due originally to Einstein) show that for a given pair of energy levels spontaneous emission can occur if and only if stimulated emission can occur. More precisely, the spontaneous emission rate when the electron is in the upper energy state is proportional to the rate at which transitions between the levels are induced when electromagnetic radiation is incident.

10.2   Superradiant Scattering

Recall the Penrose process for extracting energy from a rotating black hole. As described in 8.5, if we send a particle into the ergosphere, it is possible to split it into two fragments such that one of the fragments comes out with *greater* energy than the incident particle. (The other fragment, carrying negative energy, goes into the black hole.) It turns out that there is a wave analogue of the Penrose process. If a wave (for example, an electromagnetic wave) is sent toward a black hole, part of the wave will be absorbed by the black hole and part will "miss the black hole" and be transmitted. Under usual circumstances, the transmitted wave will have a smaller intensity than the incident wave. However, if a wave with properly chosen frequency and spatial dependence is sent toward the ergosphere of a rotating black hole, the part of the wave which is absorbed by the black hole will carry *negative* energy into the black hole (analogous to the negative energy fragment in the Penrose process for particles). The transmitted wave will have greater energy, and hence greater intensity than the incident wave. This process—first proposed independently by Ya. B. Zel'dovich and Charles Misner—is known as *superradiant scattering.*

As described above, the phenomenon of superradiant scattering can be fully understood in classical (that is, nonquantum) terms. However, one can view this process from a slightly different perspective, which suggests the occurrence of new quantum phenomena.

Recall that energy can be extracted from a rotating black hole by processes (such as the Penrose process or superradiant scattering itself) which eventually reduce the black hole to a nonrotating state. Hence, one may view a rotating black hole as an "excited state" of a black hole, just as an atom with an electron in the higher energy level is an excited state of an atom. When viewed in this manner, superradiant scattering is completely analogous to stimulated emission in atomic physics.

In view of the remark made at the end of 10.1, if one takes this analogy seriously, one is led to speculate that when quantum effects are taken into account, an analogue of "spontaneous emission" should occur in black hole physics. This analogue of spontaneous emission presumably would be spontaneous creation of particles in the vicinity of the black hole. Thus, in the early 1970s, Ya. B. Zel'dovich and A. Starobinski conjectured that particle creation must occur outside a rotating black hole and that the net effect of the particle creation would be to "spin down" the black hole to a "de-excited," nonrotating state.

10.3   Particle Creation near a Rotating Black Hole

Does particle creation in the vicinity of a black hole actually occur? Yes. Quantum field theory calculations show that pairs of particles (that is, a particle and its antiparticle) will be spontaneously created in the strong gravitational field outside a rotating black hole. All species of particles will be created (for example, electron-positron pairs, neutrino-antineutrino pairs, photon pairs, and so on), but the more massive the particle the less copiously it will be produced. One can picture the effect of this process as follows: Of each pair of particles created, one of the particles has negative energy (see 8.4 and 8.5) and negative angular momentum and goes into the black hole. The other particle carries positive energy and positive angular momentum and can escape to large distances. *Thus, a distant observer would see a flux of particles which appear to come from the black hole. As a result of this process, the black hole will lose rotational energy and eventually spin down to a nonrotating state.* Actually, we still do not have a complete understanding of what an observer stationed in the region of particle creation near the black hole would see; this is an active area of current research. However, the prediction that a distant observer would see a flux of particles appearing to come from the black hole is an unambiguous prediction of the quantum field theory calculations.

This quantum particle creation effect occurs for all rotating black holes. However, unless the black hole is extremely small the particle creation rate will be negligible. If the mass of the black hole is greater than 10^{16} gm (10 billion tons)—corresponding to a black hole with size a bit bigger than that of an atomic nucleus—no significant spin-down will occur even during the entire age of the universe from the big bang to the present. Thus, as a practical matter, this particle creation effect is of no importance for black holes which are produced by collapse processes 1 and 2 of 9.1. However, it would be important for primordial black holes of very small mass which may have been created in the early universe (process 3).

10.4   Thermal Particle Creation near a Nonrotating Black Hole

The process of particle creation near a rotating black hole was a truly remarkable theoretical discovery, but the greatest surprise was yet to come. In 1974, Stephen Hawking did a careful calculation of the quantum particle creation occurring during the gravitational collapse of a body to form a black hole. Even in the case of a non-rotating body which collapses to a Schwarzschild black hole, one would expect some particle creation to occur during the dynamic phase of the collapse because of the strong, time-varying gravitational field. In the case of a rotating body collapsing to a rotating black hole, at late times one would also expect to observe the particle creation described in the previous section. But for collapse to a Schwarzschild black hole, one expected no particle creation to occur at late times following the collapse.

However, this is not what Hawking found when he did the calculation. He found that at late times, the rate of particle "emission" to large distances does *not* drop off to zero but rather it approaches a steady, nonzero rate. Even more remarkably, this steady particle flux has precisely the character of thermal emission. By *thermal emission* we mean the following: If an ordinary body is kept in exact thermodynamic equilibrium at temperature T, it will emit particles with a characteristic spectrum that depends only on its temperature, not the detailed nature of the body. Such a body in exact equilibrium is referred to as a *black body*. In principle, all species of particles are emitted by a perfect black body but unless the temperature is extremely high (greater than a billion degrees centigrade) the emission of massive particles will be negligible and one will observe only photons, that is, electromagnetic radiation.[1] Black body thermal emission is the natural outcome of the inter-actions of the atoms in the body which result in a characteristic distribution of energy to the photons (or other particles) which does not depend on the detailed nature of the interactions. This process is of a completely different character than the process of spontaneous particle creation near a black hole. Yet, the "emission" from a Schwarzschild black hole of mass M is identical in all respects to thermal emission from a perfect black body at temperature T above absolute zero given by

$$T = \frac{hc^3}{16\pi^2 kGM},$$

where k is Boltzmann's constant. This is a truly remarkable coincidence—if indeed it is merely a coincidence rather than the consequence of some deep and fundamental idea that we do not yet fully understand.

Substituting the numerical values of the fundamental constants of nature, the above formula becomes

$$T = \frac{6 \times 10^{-8}}{M} \, °K$$

where M is expressed in units of solar masses (2×10^{33} gm) and °K denotes "degrees Kelvin." Thus, as in the case of black hole spin-

---

1. In principle, a perfect black body at temperatures lower than a billion degrees would also emit the other massless particles: the neutrino and, presumably, the "graviton" or "quantum of the gravitational field." (There does not yet exist a satisfactory quantum theory of gravitation, but it seems safe to assume that the gravitational field will display the same type of quantum wave-particle duality as the electromagnetic field.) However, the coupling of these particles to matter is so weak that in practice we would not be able to construct a perfect black body which would emit them, and even if we could it would be extremely difficult to detect them.

down discussed in the previous section, the particle creation rates are negligible unless the mass of the black hole is very much smaller than the mass of the sun. As a practical matter, the process is completely unimportant for a black hole formed by processes 1 and 2 of 9.1, but, as explored further in the next section, it would be of crucial importance for a very tiny black hole formed by process 3.

For gravitational collapse resulting in a rotating black hole, one gets a combination of the "thermal type" particle creation described above and the spin-down type particle creation described in the previous section. For a maximally rotating black hole $(J = GM^2/c)$ the emission reduces to the purely spin-down type.

10.5   Consequences: "Evaporation" of Tiny Black Holes

As described in the previous sections, particle creation in the vicinity of a black hole results in a flux of particles escaping to large distances. These particles, of course, carry positive energy. By the law of conservation of energy, the mass-energy of the black hole must decrease as this process goes on. If the black hole particle creation were only of the spin-down type described in 10.3—as was believed prior to Hawking's calculation—the process would terminate when the black hole has lost all its rotational energy; the black hole would then remain in a stable Schwarzschild state forever. However, as Hawking showed, particle creation also occurs near a Schwarzschild black hole, so a Schwarzschild black hole will also continue to lose mass.

As described in 10.4 the temperature T of the thermal emission is *inversely* proportional to the mass M of the black hole. Furthermore, the flux of energy carried by the emitted particles is roughly proportional to T^4 (that is, proportional to $1/M^4$) times the area of the event horizon of the black hole (which is proportional to M^2). Hence, as the black hole loses mass, the emission of energy from the black hole increases, and thus the rate of mass loss increases. Thus, as the mass of the black hole gets small, we have an unstable, "runaway" effect. The black hole gets "hotter and hotter," which causes M to decrease rapidly, thus making the black hole even "hotter." *Thus, in a finite time, the entire mass of the black hole will be "evaporated" away in the form of thermal particle emission.*

The lifetime τ of a black hole of mass M (measured in units of solar masses) is given by the formula

$$\tau \sim 10^{66} M^3 \text{ years.}$$

Thus, the lifetime of a black hole with mass comparable to or greater than that of the sun is enormously greater than the age of the universe. However, a primordial black hole with initial mass of about 5×10^{14} gm (500 million tons) would be undergoing the final moments of the runaway evaporation process right now.

What happens in the final moments of this process? Particle creation of all species of elementary particles occurs at an extremely high rate, resulting in an enormous burst of highly energetic particles from the vicinity of the black hole. As in the discussion of the very early universe (5.1), our present knowledge of high-energy elementary particle physics is not good enough to make detailed predictions of exactly what will emerge from the black hole. However, in any case one would get a burst of extremely high-energy gamma rays from the decay of all the unstable elementary particles which are produced. The observation of such a burst would provide a tremendous confirmation of the ideas discussed in this chapter. No such burst has yet been observed, but, of course, one does not know if *any* primordial black holes were produced in the early universe, and even if many were produced one would have to be rather lucky to be in the vicinity of one which is emitting such a burst now.

In the analysis of particle creation which we have described thus far, the particles are assumed to obey the principles of quantum theory, but the gravitational field which is causing the particle creation is treated as a classical (that is, nonquantum) entity described by the theory of general relativity. However, in the final stage of the evaporation process (Schwarzschild radius $\lesssim 10^{-33}$ cm), dimensional arguments indicate that general relativity theory breaks down and must be replaced by a quantum theory of gravitation (see the end of 4.5). Hence, one can only speculate about what happens beyond this stage. In particular, one cannot predict what will be left behind after the evaporation process is completed, although it seems reasonable to assume that one will find empty, flat spacetime at the previous site of black hole evaporation.

One further consequence of the process of black hole evaporation should be mentioned. In particle physics, *baryon* is the generic term for a proton, neutron, or other heavy particle of a similar nature. The law of conservation of baryons states that in any process the total number of baryons minus the total number of antibaryons cannot be changed. Thus, for example, a high-energy photon (baryon number 0) can get converted into a neutron-antineutron pair (total baryon number 0) but a neutron (baryon number 1) can never get converted into a pair of photons. A similar law (actually, two separate laws) applies for conservation of leptons (electrons, muons, and neutrinos). No violation of the laws of baryon and lepton conservation has ever been observed, and it is generally believed that these are absolute, fundamental laws of physics. However, the process of black hole formation and evaporation apparently violates these laws. Namely, we can form a black hole by the gravitational collapse of a body which has a large total baryon number. The gravitational field of this black hole will not retain any information about the baryon number of the body; it depends only on the total mass, angular momentum, and charge. Hence, the particle creation process cannot show any asymmetry between particles and antiparticles. Thus, with high probability the net baryon number of the particles which emerge from the black hole will be 0. If the black hole evaporates completely, one will have begun a process with large total baryon number and ended it with net baryon number 0. In a similar manner, the law of conservation of leptons can be violated.

10.6   Black Holes and Thermodynamics

Most readers are probably familiar with the following statement of the second law of thermodynamics: the total entropy of all the matter in the universe can never decrease with time. It is interesting that in black hole physics there is a very similar statement one can make as a result of the area theorem (8.6): the total horizon surface area of all black holes in the universe can never decrease with time. In fact, it turns out that a number of laws governing black hole physics can be cast in a form which makes them mathematically completely analogous to the standard laws of thermodynamics. Since the nature of the laws of black hole physics (theorems provable within general relativity) is completely different from the nature of the laws of thermodynamics (statistical laws based on the ability to average over large numbers of particles), it was generally believed that the analogy between them was purely a mathematical curiosity. However, the recent results on particle creation and black hole evaporation described above suggest the possiblity that there truly may be a deep connection between black holes and thermodynamics, as we shall now explain.

If black holes are present in the universe, the second law of thermodynamics no longer holds in the form stated above, because matter can fall into a black hole and disappear into a spacetime singularity. When it does so, the total entropy of matter in the universe decreases. On the other hand, because of the quantum, thermal emission of particles created near a black hole, the area theorem is violated; the mass and area of a black hole decrease during the evaporation process. (One of the assumptions which goes into the proof of the area theorem is not satisfied during the quantum process.) However, suppose we define the total *generalized entropy* S' by

$$S' = S + \frac{\pi}{2} \frac{kc^3}{Gh} A,$$

where S represents the total entropy of all matter in the universe *outside* black holes and A is the total event horizon area of all black holes. Then although S and A individually may decrease, it appears to be true that S' never decreases. For if we decrease S by throwing matter into a black hole, we correspondingly increase A so that S' does not decrease. On the other hand, the quantum particle creation process decreases A, but only at the price of emitting a thermal spectrum of particles, which increases S; again S' does not decrease. Thus, neither the second law of thermodynamics nor the black hole area theorem are satisfied individually, but it appears that we have a new law of physics, the "generalized second law of thermodynamics" (first conjectured by Jacob Bekenstein): the total generalized entropy S' of the universe never decreases with time.

The generalized second law is a truly remarkable law in that it involves three rather distinct fields of physics: thermodynamics, general relativity, and quantum theory. Is it merely a strange coincidence that this new law appears to be true, or is there some deep, fundamental significance behind it which we do not yet fully appreciate? At present, we cannot answer this question.

10.7   What Is the Ultimate Aim of These Investigations?

To many readers, it may seem strange that theoretical physicists are paying serious attention to the exotic particle creation and black hole evaporation ideas discussed in this chapter. After all, these ideas are based on extrapolations of general relativity and quantum theory far beyond the domain where they have been verified by observation and laboratory experiments. Furthermore, the particle creation effect is negligibly small except for very tiny black holes. It is quite possible that no such black holes exist in the universe, and even if they do it is very likely that we will never observe them. What, then, is the purpose of these investigations?

Most nonscientists seem to have the impression that the laws of physics are basically all "known," and that the work of a theoretical physicist consists solely of explaining observed phenomena in terms of these known laws. This is partly true; the known laws of physics cover an extremely wide variety of phenomena. However, each presently known law of physics has only a limited range of validity. Thus, the laws of classical mechanics developed originally by Newton accurately describe the motion of billiard balls but they are *not* valid when applied to atoms; here classical mechanics breaks down and must be replaced by quantum mechanics. The theory of quantum mechanics developed in the 1920s accurately describes atoms but does not apply to high-energy interactions of elementary particles. The principles of quantum field theory and other basic laws such as baryon conservation (see 10.5) are believed to apply to these interactions, but the detailed laws governing these interactions are not presently known. One should not be surprised if, as new phenomena are discovered, even these principles and basic laws are found to have only a limited range of validity. Furthermore, as has previously been mentioned several times in this book, one presently does not know how to merge the principles of quantum field theory with the principles of general relativity to make a satisfactory quantum theory of gravitation. The ultimate aim of many theoretical physicists is, first, to define more precisely the range of validity of the presently known laws of physics, and then, to find the new laws of physics which govern the phenomena outside this range. The discovery of such new laws is generally accompanied by a major breakthrough in our understanding of nature.

The theoretical investigation of the process of particle creation near black holes may provide physicists with a rare opportunity to make progress toward these goals. Already these investigations have suggested new limitations in previously "known" laws (10.5) and, as described in 10.6, a possible new law of physics, the generalized second law of thermodynamics. It is impossible to predict the future of physics, but it may well turn out that further understanding of the black hole will provide us with a key to further understanding of the laws of nature.

Suggestions for Further Reading

Chapters 1–3

Einstein, A. 1961. *Relativity: The special and general theory.* Translated by R. Lawson. New York: Crown Publishers.

Mermin, N. D. 1968. *Space and time in special relativity.* New York: McGraw-Hill Book Co.

Chapters 4–5

Gott, J. R. III; Gunn, J. E.; Schramm, D. N.; and Tinsley, B. M. 1976. Will the universe expand forever? *Sci. Am.* 234 (March):62–79.

Chapter 6

Schramm, D. N., and Arnett, W. D. 1975. Supernovae. *Mercury* 4 (May/June):16–22.

Chapters 7–9

Ruffini, R., and Wheeler, J. A. 1971. Introducing the black hole. *Physics Today* 24 (January):30–41.

Penrose, R. 1972. Black holes. *Sci. Am.* 226 (May):38–46.

Thorne, K. S. 1974. The search for black holes. *Sci. Am.* 231 (December):32–43.

Gursky, H., and van den Heuvel, E. P. J. 1975. X-ray emitting double stars. *Sci. Am.* 232 (March):24–35.

Chapter 10

Hawking, S. W. 1977. The quantum mechanics of black holes. *Sci. Am.* 236 (January):36–40.

Textbooks on General Relativity

Misner, C. W.; Thorne, K. S.; and Wheeler, J. A. 1973. *Gravitation*. San Francisco: W. H. Freeman & Co. (A good introduction.)

Weinberg, S. 1972. *Gravitation and cosmology*. New York: John Wiley & Sons. (An antigeometrical approach, but an excellent discussion of homogeneous, isotropic cosmology.)

Hawking, S. W., and Ellis, G. F. R. 1973. *The large scale structure of spacetime*. Cambridge: Cambridge University Press. (An excellent book but heavily mathematical and not recommended for the novice.)

Index

Accretion disk, 112
Area theorem, 102, 123–24
Asymptotic flatness, 81, 98
Axisymmetric, 86

Baryon nonconservation, 123
Bekenstein, Jacob, 124
Big bang, 47–48, 51–53
Binary X-ray sources, 111–13
Black body, 120
Black holes, 81; accretion onto, 109–10, 111–13; area theorem for, 102, 123–24; collisions of, 103–4; Cygnus X-1, 111–13; energy extraction from, 100–104; formation of, 79–82, 106–7; gedankenexperiments to destroy, 86–90; interior of, 90–91; lifetime of, 122; in a medium, 109; particle creation by, 118–23; primordial, 107–8; quantum evaporation of, 121–23; in a star cluster, 106–7, 109–10; thermodynamics of, 123–25; types of, 85–86; uniqueness theorems for, 86

Carter, Brandon, 86
Chandrasekhar, S., 70
Charged black hole, 85, 89, 90–91
Closed universe, 45, 48, 60–64
Copernican principle, 42

Cosmic censor hypothesis, 82–84, 86–90
Cosmological constant, 47
Crab nebula, 74–75
Cygnus X-1, 112

Deceleration parameter, 50, 62
Decoupling of matter and radiation, 59
Degeneracy pressure: electron, 69; neutron, 73
Density: of neutron stars, 73; of universe, 61–62
Deuterium production, 58, 63
Dicke, Robert, 32, 60
Distance measurements, 49–50

Einstein, Albert, 32–35, 47
Einstein's equation, 34–36, 46, 97
Element formation. *See* Nucleosynthesis
Energy: in general relativity, 97–99; prerelativity, 94–95; in special relativity, 95–97
Energy extraction, 100–104
Energy-momentum vector, 95
Entropy, generalized, 124
Equivalence principle, 32
Ergosphere, 99–100
Evaporation of tiny black holes, 121–23
Event, 3
Event horizon, 81; area theorem for, 102, 123–24
Expanding universe, 45–47

Galaxies, 40–41, 60